Gravetye Manor

Gravetye Manor

Or Twenty Years' Work round an Old Manor House

A facsimile of the original 1911 Victorian edition

Introduction by Tom Coward

Afterword by Elizabeth and Jeremy Hosking

Rizzoli
New York

New York · Paris · London · Milan

Introduction

Few, if any have had so great an influence on Gardening as a whole.

—Graham Stuart Thomas, from the introduction to the 1984 edition

THIS historic volume traces the birth and development of one of Britain's greatest landscapes, allowing us to follow one of the most visionary gardeners of his day as he created his masterpiece, year by year. Its writer, William Robinson, after preaching his philosophy of gardening for almost twenty years, finally had the opportunity to purchase Gravetye Manor in Sussex and put his teachings into action. From 1885 until 1907 he recorded with detail and charm his efforts in creating his ideal garden, documenting successes and failures, tenacity and determination, and most of all his complete joy for the task. It is a unique and precious archival gem and an inspiration for anyone visiting Gravetye, or with an interest in the history of gardening.

Robinson was born in 1838 in Queen's County (now County Laois), Ireland. Little is known about his early life although we may understand his hardships given the circumstances of the times: the worst years of the Great Famine began when he was nine years old and by 1852 the island of Ireland was devastated. It was in these sad and terrible times, between the ages of ten and fifteen, that Robinson started working as a garden boy on a large estate. As an apprentice gardener he would have received a strict and rigorous training; by 1861 he had become confident enough in his craft to secure a job in the Royal Botanic Society's new botanical garden in Regent's Park, London.

At this point Robinson's career seemed suddenly to take off under the mentorship of the curator, Robert Marnock. In 1863 he was awarded a grant to travel through Great Britain and Ireland to study the nation's

Interspersing naturalised herbaceous perennials amongst shrubs is an important element of Robinson's wild gardening philosophy; it embraces contrasting loose plantings with the manor's formal architecture.

Overleaf: Robinson created an outward-looking garden for this sloping site adjacent to the house to take full advantage of and celebrate views of the borrowed landscape beyond.

botanical collections, and in 1866 was elected a Fellow of the Linnean Society of London, having been proposed by, among others, Charles Darwin. The Royal Botanic Society were so pleased with Robinson's work that they offered to increase his salary substantially, but he had other ambitions and in 1867 he handed his resignation to the Garden Committee, 'Wishing to devote myself to the study of our great gardens and to the literature of horticulture for a year or two.'

He set out on a trip to France that ultimately resulted in a series of books comparing horticultural practices on the Continent to those in Great Britain, and promoting techniques such as the use of secateurs, hosepipes for watering, and planting fruit trees trained as espaliers and cordons to improve production. He went on to travel widely, taking many more trips throughout Europe and North Africa, and an extended trip across the northern United States, where he collected plants and was astonished by the magnificent trees of the Pacific Northwest. In 1871, he fulfilled his long-held dream to publish his own journal purely dedicated to horticulture, which he simply named *The Garden*. This, along with a later more affordable journal called *Gardening Illustrated*, became his pulpit, from which he could preach his vision of what gardens could and should be.

Robinson was horrified by many of the grand gardens in his era. Victorian gardens were often controlled and formal, with clipped topiary and garish bedding schemes. He saw nature as the pinnacle of beauty, so to his mind such artificial interventions were sacrilege. For Robinson, contemporary garden fashions were limited, tedious, tasteless, and unimaginative—and he wasn't afraid to force his point home with zeal. This gained him the reputation of being pugnacious and cantankerous, but what revolutionary with ideas ahead of his time doesn't ruffle a few feathers?

'There are really two styles: one straight laced, mechanical, fond of walls or bricks, or maybe gravel; fond also of geometry as the designer of wall papers excels in, often indeed of a much poorer and less graceful kind than that; fond too of squirting water to an immoderate degree,

Top: Midsummer in the flower garden, with hardy perennials achieving maximum height.

Above: Roses have always been a special part of the flower garden at Gravetye. Here Rosa *'Constance Spry' and* R. *x* odorata *'Mutabilis' harmonise with the pink Russell lupine 'Chatelaine'.*

with trees in tubs as an accompaniment, and perhaps griffins and endless plaster and stone work. The other, with true humility and right desire, though often awkward and blunderingly, accepting nature as a guide, and endeavouring to multiply her most charming features.'

Robinson advocated the use of hardy plants and the creation of herbaceous borders, meadows, and woodland gardens, always using his publications to further his views. Of the many titles he published, *The English Flower Garden* remains his best-selling, going through fifteen editions between 1883 and 1933, just two years before he died. The first part of the book outlines Robinson's thoughts on best practices, taste, and application, followed by a catalogue of plants described pithily and also illustrated by some charming black-and-white engravings: a simple but effective, user-friendly format. Possibly his most influential book, however, was *The Wild Garden*, first published in 1870 and evolving through seven issues, three editions, and several reprints during his life.

Previous pages: The sky blue of Salvia uliginosa *contrasts with the yellow of self-sown fennel flowers while rambling roses climb the manor's walls, filling the garden with fragrance.*

Above: Robinson popularised creating a matrix of shrubs and hardy perennials in his book The English Flower Garden*; the style still continues to be enjoyed in gardens all over the world.*

Robinson introduces the book himself by saying: 'I was led to think of the vast numbers of beautiful hardy plants from other countries which might be naturalised with a very slight amount of trouble, in many situations in our plantations, fields and woods a world of delightful plant beauty that we might in these ways make happy around us'. This concept of naturalising hardy exotic plants to thrive with 'careless care' in their situation was something quite new. Perhaps Robinson's creation of the Wild Garden Movement, introducing concepts of plant ecology, endurance, and sustainability remains one of his greatest legacies.

These were times of extremes; post-famine emigration had left parts of Ireland almost deserted, whereas the population of Great Britain was booming. Although there was widespread poverty, the wealth generated by the Industrial Revolution had for the first time given large numbers of people spare time and disposable income. In many ways the hobby of gardening was born along with this new middle class. Vast numbers of exciting plants were being introduced as intrepid plant hunters explored remote parts of an increasingly assessable globe. Robinson was the man who showed us how to use them.

He said, 'Now the great fact we have to face is the splendid flora of the world which has become known to us, from the stately trees of California to the flowering trees of China and the lilies of Japan. Anyone who thinks of the culture of such beautiful things can only smile at the bare idea of enjoying their life in anything but the natural and picturesque way.'

Imagine the smile Robinson would have if he could see his plants today, and how the landscape he created has matured. He facilitated his passion with a shrewd business mind and became a wealthy man. From humble and difficult beginnings he rose through the ranks of English Victorian society. In 1885, at the age of forty-seven, he was able to leave London and purchase a tumbledown Elizabethan manor in the rolling Sussex countryside: Gravetye. His achievements to this point were astonishing, but his true life's work had only just begun.

Overleaf: One of the most exciting features of the garden is the way the now-mature wild garden cascades down the slope, tumbling effortlessly into the flower garden below.

It is clear that he immediately fell in love with this gentle green countryside; he saw his main task as enhancing the landscape's natural assets. He created a new drive, taking dramatic advantage of some of the finest views in Sussex, and terraced parts of the garden to render the steeply sloping site more workable. He planted thousands of trees and hundreds of thousands of bulbs. His energy was constant yet applied with sensitive care: he took twenty years, as this very intimate volume suggests, to shape the landscape to what he considered perfection. He aimed for an outward-looking garden, where you are constantly accompanied by the view yet also sheltered from the worst of the wind that could whip through the valley. The garden contrasts wildness with formality so beautifully, and it is this balance between precision and chaos that gives constant interest. One of the most exciting aspects is the way the now-mature wild garden tumbles down the slope, cascading effortlessly into the flower garden. Robinson tells us that the inspiration for this vision came when a visiting artist was sketching and recommended removing the only clipped yew hedge in the garden: 'Why don't you give me a free line there instead of a hard, black one?' The hedge was subsequently removed and the 'evils of disfiguring trees and shrubs by clipping' were banished from the garden forever.

Over the years Robinson bought adjacent land, expanding his estate to over a thousand acres. He felt strongly that the land should be productive rather than just a park and dabbled with farming but found it a difficult business best left to the experts. He also planted thousands of trees for their future timber value as well as their beauty, and the wider estate is still used for progressive forestry research today. As his entries in this book progress, it becomes clear that these forest trees gave him some of the greatest joy. To walk today through the woodland that he started from seed in 1897 is a beautiful and humbling experience.

It is ironic that for someone who had so much influence on gardening overall, other than a garden for his publisher John Murray, Gravetye was the only garden Robinson ever actually designed or built. It is clear that

Top and above: This rolling landscape provides joy throughout the year but feels particularly dramatic when the mists of autumn and spring hang in the valley.

Overleaf: Over the years, Robinson naturalised hundreds of thousands of bulbs; daffodil Narcissus pseudonarcissus *endured and multiplied especially well, and today these descendants of his original plantings make carpets under the old handkerchief tree,* Davidia involucrata.

its construction was simply a means to an end, a stage on which to start the work that would consume the rest of his long life: 'The main fact is wholly lost sight of, viz. that the real work of the landscape gardener only begins when the garden is completed'. He finished this book in 1907 and continued to live and garden at Gravetye until his death at the grand old age of ninety-six. Despite an accident confining him to a wheelchair for the latter part of his life, he continued to care for his estate, converting a Citroën car with caterpillar tracks so he could access his meadows and tree plantings. In 1934 he was offered a knighthood by King George V, which interestingly he declined, simply replying, 'Owing to my great age and infirmity, I feel I must leave life as I entered it, and therefore decline with renewed thanks'. He died a year later, leaving behind a legacy of writing that has gone on to influence landscapes all over the world.

After Robinson's death the garden fell quickly into neglect. Then came the Second World War, and the Canadian troops stationed in the manor house understandably had little time for gardening. It wasn't until 1957, when Peter Herbert came to Gravetye and transformed the manor into the world's first country house hotel that the garden was once again

Much has been achieved at Gravetye under the custodianship of Jeremy and Elizabeth Hosking, demonstrated partly by the creation of a new restaurant designed to showcase the garden and offer a memorable dining experience.

revived. During the previous twenty-two years of neglect, the garden had become a sea of weeds and decay, but despite this, the resilience of Robinson's wild garden shone through: the structure, the tree plantings, the thousands of naturalised bulbs still spreading. For years, Herbert invested in the garden, restoring it to its former beauty and preserving Robinson's legacy. Possibly one of his toughest challenges was the 1987 hurricane, which destroyed Gravetye's tree line. Much was lost on that terrible night, although it did give the opportunity for replanting which, in a mature garden, is a valuable thing. With the help and advice of many accomplished gardeners, Peter Herbert made an especially wonderful job of this, leaving us a healthy mix of tree generations throughout the garden. For nearly fifty years he cared for Gravetye, establishing it as one of the finest hotels and kitchens in the world and giving the garden purpose for the future. If Gravetye owes its creation to William Robinson, it owes its salvation to Peter Herbert.

Old historic gardens like Gravetye demand passionate care and significant investment to keep them alive. Even one season of neglect can mean several subsequent years of cleanup, as a wild garden can quickly descend into wilderness. Business had slowed at the hotel in the years leading up to Herbert's retirement in 2005, and in the years following the garden began to suffer. Many did what they could for the place, although the resources for proper maintenance simply weren't available until Jeremy and Elizabeth Hosking took ownership of the hotel in 2010. Their determination has been to create a viable revenue stream for the property by restoring the garden and renovating the hotel, enabling the care of this unique piece of heritage for future generations. Much has been achieved under their leadership: deer fences built, glasshouses restored, borders replanted. Perhaps the most notable development has been the beautiful new dining room and restaurant, built in 2018, designed with a view that showcases the garden. With a third generation of dedicated custodians, this beautiful garden has been resurrected to its full glory once again and faces a bright future.

THE GARDEN TODAY

When I first accepted the job as head gardener it was an exciting yet slightly daunting challenge. The flower borders had become a sea of weeds needing complete replanting and much of the garden's infrastructure required rebuilding. The heritage of this beautiful landscape had to be preserved while renovating the garden for its modern purpose as a hotel and restaurant. In the flower garden we have allowed ourselves to be quite free with our plantings as they were also constantly changing in Robinson's day. What is important is that we continue to apply his principles. Developing an appropriate planting style has taken time, planning, and research. As well as his many books, possibly the most useful archive has been paintings of the garden. These show the evolution of his planting style and capture the atmosphere he created like nothing else.

The flower borders at Gravetye continue to champion the hardy perennial as Robinson originally intended, and we still find space for his old favourites such as his hybrid tea roses, carnations and 'starworts' (asters). But we never wanted the garden to become a museum and felt that it was only right to continue to move the garden forward in his spirit. As a hotel garden it is important to have as long a season of interest as possible to entertain our guests at any time of year. To achieve this we sometimes deviate from Robinson's original plans by using successional plantings of annuals, bulbs, and late-flowering tender plants amongst the shrubs, roses, and hardy perennials. This adds an extra element of interest throughout the year and if he were able to see it now, I hope he would approve.

The flower borders are ephemeral, taking constant work and always changing. But it is the wild garden elements throughout Gravetye that are, in terms of heritage, most important and where much of Robinson's original work, detailed in this book, can still be seen. There is no doubt

Top: William Robinson was especially passionate about heathers, and introduced concepts for using them as garden plants in a late chapter of this book. Here ancient specimens of Erica carnea *mix with* Euphorbia characias *to create a stunning display in the wild garden in early spring.*

Above: Asters, or starworts as Robinson liked to call them, were another favourite. We find Aster pyrenaeus *'Lutetia' quite effective for fall colour at the front of this border.*

Previous pages: Only one plant of Wisteria floribunda *'Shiro Noda' survived from Robinson's original pergola. This was restored in 2011, and today creates one of the seasonal highlights of the garden.*

Top: Camassia leichtlinii *subsp.* suksdorfii *(Caerulea Group) is one of the best bulbs to combine with apple blossoms in the orchard meadow.*

Above: Leucojum aestivum *'Gravetye Giant' is a vigorous and floriferous bulb that naturalises well but performs best in damp soils.*

that he was a perfectionist and each considered mark he made upon the landscape was intended to last. His philosophy 'In all country work true economy is the enduring way' applied to everything he did and was an important element of his 'wild gardening'. Periods of care and neglect, storms and droughts, pests and diseases have all left their mark over the past century and it is fascinating to see how Robinson's plantings have endured.

Perhaps one of his most successful planting projects was in his meadows, where he naturalised hundreds of thousands of bulbs that continue to thrive today. These meadows run throughout Gravetye, repeating through the landscape and occupying the greatest area of the garden. They start flowering early in the spring, first with the snowdrops, followed by little *Crocus tommasinianus* in the coldest months. Then come carpets of daffodils in March, with the charming sky-blue *Scilla bifolia*. Our own *Leucojum aestivum* 'Gravetye Giant' colonises the damper areas whilst the beautiful pale lavender blue *Anemone nemorosa* 'Robinsoniana' makes carpets in shady spots. These plants have truly naturalised, happily spreading and growing and running free. It is likely that in another century these meadows will be even more beautiful, a reminder that a bulb planted correctly can be even more enduring than a tree. As well as the bulbs, these meadows are home to countless native plants and insects, making them havens of biodiversity.

Although much of Robinson's original bulb plantings have survived, some had been lost, and over the years and as part of our work in caring for the landscape we have taken the opportunity to restore and develop these. Each year we have added thousands of bulbs—daffodils and snowdrops, wild tulips, camas, lilies, and many, many more. Care must be taken when planting in to such a considered and sensitive landscape because if the wrong kinds of bulbs are planted they can quickly look very out of place. Most of our inspiration for the bulbs selected has come from the pages of this book, and of course *The Wild Garden*.

But the plants will always teach us better than any book, and the study of meadow gardening on other sites has been essential in our work of restoring and developing Robinson's original plantings. There are many examples to draw inspiration from, in particular the neighbouring garden of High Beeches that was so deeply influenced by Robinson. Also my experience working at Great Dixter, a garden thirty-five miles east from Gravetye, also on the Sussex Weald. There Robinson's ideas on meadow gardening continued to be developed by Daisy Lloyd and her son Christopher, refining this form of wild gardening to a fine art. The years I spent helping care for them were formative, and bringing this experience back to the meadows at Gravetye has been especially rewarding. Planting tens of thousands of bulbs in an autumn is a slow and monotonous process but, in the long term, watching these plants establish and spread is the most exciting thing in the world. We have also started many bulbous plants from seed, both directly sown into the meadow and into pots in the nursery. Nursery propagation gives us the opportunity to raise large numbers of plants that are quite expensive and hard to find commercially or otherwise. It can take up to five years from seed to flowering for a lot of bulbous plants, but it is well worth the wait and a relatively short time in the overall context. Also these plants have the best chance of establishing well and seeding themselves for future generations.

Another part of Robinson's wild garden legacy was his contribution to the development of the woodland garden, shown in his beautiful essay at the end of this book. Although several storms have been cruel to our trees, many magnificent specimens from Robinson's time still survive at Gravetye. It is difficult to pick favourites, but the glorious pink *Magnolia campbellii* is always a special experience when it is in full flower. Planted in 1897, it would have taken around thirty years to start flowering and only now is it at its best. As well as veteran trees, surviving plants of Robinson's garden include antique heathers and heaths, hardy azaleas and rhododendrons, accompanied by the subsequent plantings of further generations.

Previous pages: The spring awakening of hundreds of thousands of bulbs.

Top: A fine specimen of Magnolia campbellii, *one of the earliest and most beautiful of all magnolias introduced to the garden.*

Above: Rhododendron hybrids on the edge of Robinson's woodland garden.

Cultivating exotic and native woodland plants in the shade of his trees is possibly one of the most challenging, yet rewarding, forms of wild gardening. Our job as gardeners is to ensure we also leave a succession of young trees and plants to contribute to this continued endurance. Selecting the best tree for the right spot is a lifelong study and luckily we are blessed with many great woodland gardens nearby for inspiration. High Beeches, Leonardslee, Nymans, Borde Hill, and Wakehurst Place are some of the finest such gardens in the Southeast of England, all less than twenty miles from Gravetye. Their development has been directly influenced by Robinson, and by studying these landscapes we can bring ideas back to the care plan of the woodland garden at Gravetye. Much has moved on since Robinson first inspired the woodland garden, most notably the vast numbers of new introductions and the subsequent breeding of wonderful garden plants. I have no doubt that he would embrace the wealth of plants

Previous pages: Azalea *'Irene Koster' is one of the most reliable and effective deciduous azaleas; it works well in the wild garden with a groundcover of hardy perennials, bulbs, and self-sowing annuals.*

Above: It took twenty years' consideration and three attempted locations before Robinson found the perfect place for his massive kitchen garden. He ultimately enclosed it with an elliptical sandstone wall to create the a desirable microclimate for food production.

we have today. Introducing a new generation of trees after his original woodland garden ideals has been an especially exciting task. Woodland gardening has gone on to become an important style of gardening in its own right, with a global appeal.

One of Gravetye's many charms is its diversity, which provides constant interest through the year. The landscape and structure of the garden show to beautiful advantage in the bleakness of winter, and I find it especially wonderful when the snowdrops approach their peak in February. Spring is magnificent in any garden but Robinson's bulb meadows, along with countless blossoming trees, really are a sight to behold. Another equally impressive highlight comes when our stunning white wisteria pergola blooms at the beginning of June, filling the flower garden with fragrance. Through summer, the borders of hardy perennials and roses are at their finest and as they mature towards late summer, asters, dahlias, and sages are in their full glory; then autumn colours grace the landscape.

Possibly one of the most magnificent features at Gravetye is the grand elliptical walled kitchen garden that sits at the highest point of the garden. This was finished in 1900, but it took Robinson seventeen years of consideration and many moves for to decide on the perfect spot in which to grow food. One of the best aspects of my job is to continue growing crops here, with my friend Helena Lawson and a small team of gardeners. Much of what we do is unchanged from when the garden was first built, as we still grow produce in traditional ways, and all of it is consumed in Gravetye Manor. The only difference from Robinson's day is that instead of the garden feeding a Victorian estate, we are supplying a Michelin-star restaurant, where our head chef, George Blogg prepares some of the best food in the world.

—Tom Coward, Head Gardener

Previous pages: Fruit and vegetables, flowers grown for cutting, and beneficial companion plants make up an attractive matrix in the walled kitchen garden.

Above: High summer in the kitchen garden, as tomatoes ripen and sweet corn flowers sway over the foliage of pumpkins and squashes.

Right: The kitchen garden is tasked with producing edibles twelve months of the year to supply the chef's seasonal menus. Red cabbage is sculptural, as well as a particularly delicious crop for winter harvesting.

Previous pages: Edible flowers have become an important element in many of the dishes at Gravetye, offering unique flavours and beautifying each plate. Here evening primrose, nasturtium, borage, calendula, and fennel flowers are all ready for the kitchen.

Above: An old tea rose in the shade of cedar trees.

Gravetye Manor

Or Twenty Years' Work round an Old Manor House

Being an Abstract from the Tree and Garden Book of Gravetye Manor Sussex, kept by the owner, WILLIAM ROBINSON, Author of 'The English Flower Garden,' 'The Wild Garden,' 'The Parks and Gardens of Paris,' 'God's Acre Beautiful,' 'The Garden Beautiful' 'Home Woods and Home Landscape'; Founder and Editor of 'The Garden,' Vols. 1 to 56; 'Gardening Illustrated' Vols. 66; 'Woods and Forests,' Vols. 2; 'Flora and Sylva,' Vols. 3. Thirty-five years Garden and Woodland Editor of 'The Field'

Introduction by Graham Stuart Thomas
with a contribution by Will Ingwersen

Afterword by Peter Herbert

WARREN'S WOOD

Bibliography

William Robinson 1838–1935

It is truly remarkable just how productive Robinson was as an author. Counting all editions and issues, he published more than seventy books over his long career*. He officially retired in 1919, although he continued to work on several more books, including a special-edition print of *The Wild Garden* in 1928. Even in 1932, at the age of 94 and just two years before his death, he sent his publisher additional script intended for an eighth edition of *The Wild Garden*. This was unfortunately not printed until 1994, but it shows his complete devotion to his craft throughout his life and the sharpness of his mind even at such a great age. It has been commented that it would be almost impossible to compile a complete and comprehensive bibliography of his work because of the numerous periodicals that he founded himself. To all of theses he contributed countless notes, articles and editorials which were often unsigned. This, however, is a list of his major titles and some of the more recent biographies.

1868 *Gleanings from French Gardens*: comprising an account of such features of French Horticulture as are worthy of adoption in British Gardens

1869 *The Parks, Promenades and Gardens of Paris* described and considered in relation to the wants of our own cities and of public and private gardens

1870 *Alpine Flowers for English Gardens*

Mushroom Culture: its extension and improvement

The Wild Garden: or our groves and shrubberies made beautiful by the naturalisation of Hardy exotic plants. With a chapter on British wild flowers. (This popular book has had several reprints in recent years, most notably a new illustrated edition with photographs and notes by Charles Nelson in 2010 and the expanded edition, with new chapters and photographs by Rick Darke in 2009.)

1871 *A Catalogue of Hardy Perennials*

The Subtropical Garden: or beauty of form in the flower garden

Founding of *The Garden*

1878 Founding of *La Semaine Francaise*

1879 Founding of *Gardening Illustrated*

1880 *God's Acre Beautiful*: of the Cemeteries of the future

1881 *Asparagus Culture*: the Best Methods Employed in England and France

Founding of the *Garden Annual, Almanack & Address Book*

1882 Founding of *Farm and Home*

1883 *The English Flower Garden*: Style, position and arrangement; followed by a description, alphabetically arranged, of all the plants best suited for each (The most recent print of this, his most successful book, was a reproduction of the fifteenth edition in 1984, with a foreword by Henry Mitchell and introduction by Deborah Nevins.)

Founding of *Woods and Forests*

1889 *Cremation and Urn Burial*: or the Cemeteries of the future

1892 *Garden Design and Architects' Gardens*

Founding of *Cottage Gardening*

1903 Founding of *Flora and Sylva*

1907 *The Garden Beautiful, Home Woods and Home Landscape*

1911 *Gravetye Manor: or, Twenty Years' Work round an Old Manor House* (Also reproduced in 1984, with introduction by Graham Stuart Thomas and afterword by Peter Herbert.)

1912 *The Virgin's Bower*: Clematis, Climbing Kinds and their Culture at Gravetye Manor (Reproduced by the British Clematis Society, 1996 with foreword by Peter Herbert and introduction by Mary K. Toomey.)

1914 "Civil War and Party Lawyers" (Pamphlet)

1917 *My Wood Fires and Their Story*

1924 *Wood Fires for the Country House and Cottage*

1935 *Clematis*: The Large and Small Flowered, Their Cultivation in the Open Air, by Ernest Markham, head gardener at Gravetye, with introduction by William Robinson

* Charles Nelson, introduction to *The Wild Garden*, 2010.

BIOGRAPHIES OF WILLIAM ROBINSON

Some Nineteenth Century Gardeners. Geoffrey Taylor. Ancoor Press, 1951.

William Robinson 1838–1935: Father of the English Flower Garden. Mea Allan. Ebenezer Baylis & Son, 1982.

William Robinson: The Wild Gardener. Richard Bisgrove. Frances Lincoln, 2008.

Glossary

Graham Stuart Thomas

Abele = *Populus alba*
Acacia = *Robinia pseudacacia*
Aconites = *Aconitum*
– Winter = *Eranthis hiemalis*
Alder = *Alnus glutinosa*
Almond = *Prunus dulcis (P. communis)*
Anemone, Appenine = *Anemone appenina*
Ash = *Fraxinus excelsior*
Ash, Mountain = *Sorbus aucuparia*
Aucubus = *Aucuba japonica*
Bamboo, palmate = *Sasa palmata*
Barberry = *Berberis vulgaris*
– Evergreen = *Mahonia aquifolium*
Bay Laurel = *Laurus nobilis*
Birch, canoe = *Betula papyrifera*
Bloodroot = *Sanguinaria canadensis*
Box = *Buxus sempervirens*
Bramble, cut-leaved = *Blackberry Parsley Leaved*
Broom = *Sarothamnus* or *Cytisus scoparius*
Broom, Butcher's = *Ruscus aculeatus*
Broom, Spanish = *Spartium junceum*
Buckthorn, Sea = *Hippophaë rhamnoides*
Bulrush = probably *Typha latifolia*
Cedar, Red = *Juniperus virginiana*
Cherry = *Prunus avium*
– Bird = *Prunus padus*
– Plum = *Prunus cerasifera*
Chestnut = *Castanea sativa*
Clematis, wild = *Clematis vitalba*
Daffodils, Poet = *Narcissus poeticus*
Dock, Water = *Rumex hydrolapathum*
Dogwood, Siberian = *Cornus alba*
Dyer's Weed = *Genista tinctoria*
Elm, Wych = *Ulmus glabra*
Erica codonodes = *Erica lusitanica*
Euonymus, evergreen = *Euonymus japonicus*
Fern, Royal = *Osmunda regalis*
Fir, Calabrian = *Pinus nigra* var. *maritima*
– , Douglas = *Pseudotsuga menziesii*
– , Scotch = *Pinus sylvestris*
– , Silver = *Abies*
Flag, Sweet = *Acorus calamus*
Forget-me-not, Water = *Myosotis scorpioides (M. palustris)*
Foxglove = *Digitalis purpurea*
Furze or Gorse = *Ulex*
Gale, Sweet = *Myrica gale*
Globeflower = *Trollius europaeus*
Goat's Beard = *Aruncus dioicus*
Grass, Pampas = *Cortaderia selloana*
Hazel = *Corylus avellana*
Heath, Alpine Forest = *Erica arborea etc.*
– Connemara = *Daboecia cantabrica*
– Heather = *Erica, Calluna*
– Hybrid = *Erica × darleyenesis*
– Mediterranean = *Erica erigena (E. mediterranea)*
Hemlock, Carolina = *Tsuga caroliniana*
Holly = *Ilex aquifolium*
Honeysuckles = *Lonicera caprifolium* and *L. periclymenum*
Hornbeam = *Carpinus betulus*
Ivy = *Hedera helix*
Juniper = *Juniperus communis*
Knotworts, Knotweeds = *Polygonum*
Larch = *Larix decidua*
Laurel, Cherry = *Prunus laurocerasus*
Spurge = *Daphne laureola*
Laurustinus = *Viburnum tinus*
Lily, American = *Lilium pardalinum* etc.
– American Swamp = *Lilium superbum*
– Belladonna = *Amaryllis belladonna*
– of the Valley = *Convallaria majalis*
– St. Bruno's = *Paradisea liliastrum*

- Tiger = *Lilium tigrinum*
- Turk's Cap = *Lilium chalcedonicum, L. martagon L. superbum*

Locust Tree = *Robinia pseudacacia*
Loosestrife = *Lythrum* and *Lysimachia*
Maple = *Acer campestre*
Maidenhair Tree = *Ginkgo biloba*
Meadowsweet = *Filipendula ulmaria*
Mirabelle Prune = *Prunus cerasifera*
Moccasin Flower = *Cypripedium*
Moon Daisy = *Chrysanthemum uliginosum*
Moneywort = *Lysimachia nummularia*
Narcissus, Bayonne = *Narcissus pallidiflorus*
Nuphar, American Yellow = *Nuphar advena*
Oak, Evergreen = *Quercus ilex*
Pampas Grass = *Cortaderia selloana*
Pine, Austrian = *Pinus nigra*

- Corsican = *Pinus nigra* var. *maritima*
- Pitch = *Pinus palustris*
- Riga = *Pinus sylvestris,* a Baltic strain
- Weymouth = *Pinus strobus*

Plane = *Platanus* × *hispanica*
Poplar, Black Italian = *Populus* × *canadensis* 'Serotina'
Canadian = *Populus* × *canadensis*
Quick (thorn) = *Crataegus monogyna, C. oxyacantha*
Rose, Alpine = *Rhododendron ferrugineum*

- American Marsh = *Rosa palustris*
- Dog = *Rosa canina*
- Tea = forerunners of Hybrid Tea roses

Rush, Flowering = *Butomus umbellatus*
Savin = *Juniperus sabina*
Sloe = *Prunus spinosa*
Snake's Heads = *Fritillaria meleagris*
Snow Glories = *Chionodoxa*
Solomon's Seal = *Polygonatum*
Spindle Tree = *Euonymus europaeus*
Spruce, Black = *Picea mariana*

- Blue = *Picea pungens*
- Hemlock = *Tsuga*
- Norway = *Picea abies*
- Numidian = *Abies numidica*
- Sitka = *Picea sitchensis*
- White = *Picea glauca*

Sycamore = *Acer pseudoplatanus*
Starwort = *Aster novi-belgii (Michaelmas Daisy)*
Sumach, Venetian = *Cotinus coggygria*
Sweetbrier = *Rosa rubiginosa*
Teazle, Fuller's = *Dipsacus fullonum*
Thorn, Cockspur = *Crataegus crus-galli*
Violet, Dog's Tooth = *Erythronium*
Wellingtonia = *Sequoiadendrum giganteum*
Willow, Cardinal = Allied to *Salix alba* 'Britzensis'

- Herb = *Epilobium*
- White or Grey = *Salix alba*
- Yellow = *Salix vitellina*

Yew = *Taxus baccata*

Preface

OFTEN, when seeing places of interest, I have thought it would be well if one could know the ages of trees and plantations. So in possession of a place of my own I resolved to keep a record of all the work done in the garden and woodland, and, as a rule, the reason for it. Fate gave me a piece of land in which all had to be done—every field, every wood, every cottage, every farmhouse, every road, and every fence had to be overhauled. The roads, which had sufficed for the Tudor days when people rode horses more than they do now, were in grade quite useless for modern use. The books I kept included building and many works not of general interest, and what is given here concerns garden design and planting and landscape work only. I thought that the record might be of use to others who are striving to get all the beauty that is possible out of a neglected country place or one in being formed anew. Everything was done with regard to landscape beauty, whether roadside fence, plantation, covert—all was done with that in view. The sloping ground round the house made terracing necessary, and, like all work of that kind, it gave rise to much labour, but it was a mere fraction of what arose when all the garden ground round the house was finished. In the fruitless writing of the last few years about formalities and informalities in relation to garden design, the main fact is wholly lost sight of, viz. that the real work of the landscape gardener only begins when the garden itself is completed. It is most important to think of this in all garden design or planting.

The greatest and the most lasting improvement ever made in English gardens was the breaking away from the old idea of gardens of barrack-yard uniformity, such as we see several illustrated in many old books. It was this happy change which struck the foreigners most, and was recognized by them as on the whole the most artistic thing ever done by English gardeners. In the old days there were so few plants that architects made gardens of stones and a few clipped trees with geometrical tracery, until by great English writers the folly of it was shown. It is to that we owe any beauty that there is now in English gardens —in our Dropmores, Bictons, Tregothnans, and many other places in England and in Ireland as well. There is not a garden in any county in England which does not owe any dignity or beauty it possesses to the freedom and grace of landscape gardening. Writers on this theme deal too much in words and too little in facts, and that too with beautiful gardens before them. On the Continent, if one goes into a beautiful picturesque garden in the neighbourhood of Vienna, or in one of the valleys of France, the chances are that he will find it known as the 'English garden'. Before the English invented and carried out landscape gardening the foreigner, like ourselves, was content with the mason's idea of a garden.

Now the great fact we have to face is the splendid flora of the world which has become known to us, from the stately trees of California to the flowering trees of China and the Lilies of Japan. Any one who thinks of the culture of such beautiful things can only smile at the bare idea of enjoying their life in anything but the natural and picturesque way. Take a beautiful river like the Severn, or the Barrow in Ireland, where people are happy in having gardens on the banks of such a river how much it means in planting. The knowledge of trees and shrubs would come in there effectively; and all matters such as composition, breadth, repose, and the many things which have to be considered and to which the term 'landscape gardening' is truly applicable. Even in places of only moderate extent the planting of trees from all the northern countries of the world deserves much thought, and without a knowledge of them it is hopeless to expect the best results. The noblest trees we know were unknown to us until recent times. They are now coming into use not merely as ornamental things but as forest trees, and in view of their value no one should attempt to practise the art of landscape gardening, or planting, without knowing them both in the wild and in the cultivated state. The facilities for travel now are such that there need be no difficulty in acquiring this knowledge. In planting such trees in the many conditions of our country of river, estuary or lake, or even in ground without these advantages, we have almost the same problems to consider as landscape painters often have as regards composition and other effects.

Having seen for many years the evils of disfiguring trees and shrubs by clipping, too evident in many gardens, both public and private, I naturally did not follow a plan so evil in its results. It is a common cause of hard and ugly lines in gardens, and usually so needless as well. We have so many beautiful Evergreens, from Bay to Holly and Cypress, to give us fine backgrounds, that it is often easy to get all we want in that way without effacing natural form. This foolish waste of energy to produce ugliness is often the result of planting shrubs and trees in the wrong places—a practice which any one who knows his work should not do. The Yew is the most frequent victim of this evil practice and is cut into various ugly shapes and hard lines. Nothing is more miserable for the gardener, or uglier in the landscape, than a garden laid out with clipped Yews. The evil did not occur to the same extent in old places, because then the Yew was often planted for shelter; but planted as part of modern garden design its effect is often bad. Although dissenting from this practice, I had some respect for it in a modest degree, as in some cases it is useful as a dividing line; and I put a small hedge between two flower gardens on different levels where I wanted an emphatic division. One day, Mr. Mark Fisher, the landscape painter, was sketching the flower garden, and this

Yew hedge formed the extreme end of the picture. He said to me, 'Why don't you give me a free line there instead of a hard, black one?' The remark struck me very much. Next Autumn I took away the hedge of Yews and planted the noble rose, Bouquet d'Or, mixed with the claret-coloured Clematis, and both have formed a charming dividing line for years. The lesson I never forgot; we abolished the shears and clipped no more. We have such riches now in the lovely climbing shrubs from parts of the world hitherto not much explored for plants and varied Evergreens, that to any one who studies our wealth in this way, the forming of dividing lines and shelters should be easy without the aid of the shears. Without naming the most grotesque examples of tree mutilation in England, it is clear that much beauty is lost in our gardens by the stupid and ignorant practice of cutting trees into unnatural shapes. And it is not only in its results in ugliness, but also in the waste of labour it entails—in many cases acres of Laurels are cut into hard level plateaux every year—that some of the worst results of the practice are seen in important public gardens as well as in private places.

W. R.

OLD SUSSEX FOOT-SCRAPER

Contents

GARDEN AND LANDSCAPE

CONTENTS

1895

1896

1897–8

1899

1900–1

CONTENTS

List of Illustrations

GRAVETYE MANOR

GARDEN AND LANDSCAPE

1885

COMMENCED on August 24 to form this above the house and to the N. of it, raising the level of the E. end a good deal. Removing broad Spruce hedges, a rosery, fernery, 'rockery'. The effect of it when finished, good. *Tennis Lawn N. of House.**

Cleared in November the grassy bank above farm-yard and near and below the bailiff's cottage, taking away many small trees of various kinds, planted, no doubt, to hide view of farm-yard. Some Oak, Ash, and other trees also removed, to give space to the larger Oaks, and allow them alone to grace the slope, and to give a view into the valley. The removal of one tree near the centre opened a noble view. Stood at the E. end of the opening on the drive, watching the tree fall, and was surprised to see the picture that came into view as its head (one of no great size) came to the ground. *Grass Slope above Farm-yard.*

It is hoped that the Oaks left here, now with all the space they require, will go on increasing in size, and form a picturesque group.

Determined to do away with the old kitchen garden, which was right against the house, and had a poor effect. Left the older Apple-trees, and planted the vacant parts with *Site of old Kitchen Garden.*

---

* NOTE, 1894. *Though done with care, and by men trained to the work, the soil in the 'made' part sunk nearly a foot in five years, and the surface had to be raised and levelled again, showing the need of all ground moving being done with great care.*

1885 groups of the best Apples, to form, in years to come, a pretty Orchard on the rising ground.

Site of old Kitchen Garden. Many iron trellises, a wall running E. from house and supporting the old kitchen garden, 7 or 8 ft. above level of approach, were removed, and the ground lowered and graded with care to the natural line of the surface near.

West Garden. The view of the old Yews being marred by a row of Cedars between them and the house, the Cedars were cut down, with an excellent result. The garden was levelled at the same time and simple walks formed. The bank between this west garden and the play-lawn was high and awkward, with a line of poor Cedars along it. The surface was changed and lowered, carrying the spare soil round to the W. end of the garden, and so forming a gently sloping bank all round it. A great quantity of soil was moved here, and new steps formed to reach the bank above in the easiest way.

East Approach. On the level space before the eastern wing was a mass of rock-work (so called) of the ghastly order. It touched the NE. corner of the house, and extended in an arc of a circle towards the W. wall of the old kitchen garden, and within view of the E. windows. Consisted of stone and small rocks, stuck in promiscuously, with great Ash pollards (sawn in lengths) reared on end amongst them, the effect being hideous. The whole was abolished in favour of a sloping bank, with stone steps at its W. end, to connect the lower and higher levels.

Trees condemned. All the Wellingtonias were cut down as being at once ugly and unsuited to the climate. Also many Spruce, which rarely ever does well in this part of England.

Scotch Firs. The groups and belts of Scotch Firs to the SE. of the house freed from a number of other trees and some Spruce, which prevented them thriving. They formed part of the usual 'mixed' plantation; but seeing these firs promised to give, in time, the best effect, removed all the other trees. The result the first year was good, but the true effect I sought can be best seen when the trees get old. 1885

Wild Gardening. Many Daffodils of various kinds were planted in the wild meadows in November: common English sorts by lower end of lake, Poets on the lawn near the Oaks to the right as one goes down by Smuggler's Lane, Poets along the streamlet in the moat meadow, and Tenby Daffodil on the slope of Sweetbrier meadow, where it falls to the water. All were planted as far as possible in free 'natural' looking colonies, the centre somewhat closely massed with outlying fringes and groups.

About the same time thousands of bulbs of Star Narcissus (*N. Incomparabilis*) and its varieties were planted in groups and colonies among the grass on both sides of the drive, and on grass slope above farm-yard. To plant them the turf was turned back by cutting the sods with one side left as a hinge—three cuts made with the spade, thus △, and the sod lifted up and turned over. From three to five bulbs were set in the bottom or sides of each cavity, and the turf then turned back and trampled down firmly.

Wild Tulip (*T. Sylvestris*) bulbs were planted in one large colony near the Oaks on the upper (N.) side of the Sweetbrier meadow. These were sent me by Edouard André, from Touraine.

Rabbits. The Estate was overrun by these pests when I came—fields half bare, and in Bushy Wood six acres of underwood eaten clean down. Over a thousand were killed during three months of the autumn and winter, so that for generations Gravetye has not been so clear of them. Notwithstanding this, some Yellow Willows (a good set sent from Eastnor) which were planted at the end of the lake were gnawed off by rabbits a few days after they were put in. I shall probably have to wire in the whole place, being surrounded by woods in which rabbits abound.

Stables Removed. Decided to abolish the old stable which stood opposite the south-east corner of the house (see Photograph Book). Such

1885 — *Stables Removed.* a building so very near the house seemed to me in various ways objectionable, and the house has gained in freedom and dignity by the removal. New stables will be built beyond the road above the old kitchen garden and newly-made orchard, where they will be concealed from the house by trees and yet sufficiently near.

Old Garden Walls. Except by readjusting the steps, no alteration was made in the beautifully clothed old wall along the south side of the gardens: but the high wall on the east side of the little front garden, and extending southwards from the house, was built at this time as a necessary protection to the garden as well as effectually to cut off the garden from the approach side of the house.

SANDSTONE ROCKS, MILL PLACE

1886

Flower Garden. Commenced on March 25 to form the simple beds on the grass in the west garden, and to plant Tea Roses in the beds and against the walls.

New Orchard East of House. This is on the site of the old kitchen garden. Owing to the heavy ground work involved in getting the very awkward

MOAT HOUSE FROM SMUGGLER'S LANE

ground into one 'natural fall' towards the Drive, the planting could not be done till late, and I put in all the Apple-trees on April 2. I chose for this select plantation kinds I knew to have high quality—not merely the most popular kinds—and have planted a dozen of each, so as to secure a good 'gathering' of it.

1886 The ground of this new orchard having been deeply trenched, the trees will do well without keeping the ground under them cultivated for a few years, in the Kentish way.

Scotch Firs at House. The groups on the east side of the house were now freed from all remains of the old stables, the rampant Evergreens and other trees, and also from poor specimens of their own kind, so that their effect might now be seen. These Firs were planted about twenty-five years from this date (1886).

Lawn S. of House. The long grass 'terrace' which was formed of earth two years before my coming was now removed, and the ground restored to its natural level or form. The terrace ran across from side to side below the Elms south of the house, and was very ugly. Its removal occupied eight men with the necessary horse-labour for about two months.

Old Road to House abolished. The old road leading from Turner's Hill to the house having been for some years closed as a carriage way, decided to close it, diverting the right of way, if possible, by the moat. The road down to the Moat Cottage was made as good as a farm road.

Carnations. Formed a large collection of 'hardy' border kinds, 'self-coloured,' over 500 plants coming from France. Not many of those plants survived the winter owing to rabbits and frost.

Moccasin Flower. Planted small groups in bog moss in pond (old pit) in Bushy Wood (November 10): sent me by Mr. Elliot of Philadelphia.*

Reservoir of Water for House Supply. This, which is built in the high ground north of house and within the limits of the new kitchen garden, was emptied and cleaned on February 24. It has a capacity of a little over 12,000 gallons, and is well built and finished on the inside with Portland cement. The supply is from a spring near.

Sunk Fence to South Lawn. Formed in September: to secure freedom of view from south of house towards the lower lake and also across top of Sweetbrier Mead to the Rocks Farm: it was made of Rowfant bricks and cost about £325.

---

* NOTE. *Never could find a trace of the plants since.*

1887

Removed in February; the soil scattered over the south lawn and opposite field. Effect much better as the island was prim and too artificial. Lake cleaned out at the same time. *Island in Upper Lake.*

The lower lake was cleaned out in 1885 in all the upper parts, and also near beach in 'sands' near lower island: again cleaned out in 1887.

Changed to present form, outline greatly simplified and fences near improved: poor, isolated, and stunted trees removed. *Paddock at Home Farm.*

Grubbed, and the ground, which was very irregular, thrown into a more easy form without obliterating the traces of the continuation of Smuggler's Lane. Old road to house dug up and sides levelled, leaving it for a pathway through new Evergreen grove. Planted with hardy Evergreens in autumn of same year. *Copse West of House planted.*

A group of these silvery Willows planted at lower end of upper lake to hide stiff bank from the house: the stews between the ponds repaired, and some wall formed in spring. *Bedford and Huntingdon Willows.*

Put all round garden and pleasure ground with a view of keeping rabbits out, concealed in various ways so as not to come into the view: by its aid and the barriers afforded by new sunk fence and water hoped to keep the contents of the gardens from injury by rabbits *. *Iron and Wire Rabbit-proof Fence.*

---

* NOTE, 1894. *With care and watchfulness this saved the garden from any injury from rabbits for years, though the woods quite near were swarming.*

1887 — *Yew Grove N. and SE. of Orchard.* Soon after the new stables were built determined to do away with the old state of things here and raise a bank of earth N. and E. of the orchard, getting the soil mostly at the quarry where the stone for the new stable was dug, and some out of the foundations, cisterns, and 'cuttings' at the stable. It was poor clay with a spit of good soil over all, and was formed in time to allow of the yews being collected and planted by Christmas. The yews were mostly got in the Surrey nurseries—very fine plants from 3 to 8 ft. high. This great bank of yew is intended to hide the stable from the house, and also to form an evergreen shelter N. and E. of the orchard, and also to shelter the house and its approaches from the east.

The Moat House. On June 1, commenced repairing the Moat House, an old farm-house of great age, which had a moat, now dry. The house appears to have been built or rebuilt at three different periods, judging from the state of repair and the manner in which the oak timbers had been wrought. It is built in three parts, the centre part, about 25 ft. by 21 ft., being the most modern, and the two ends, about 12 ft. by 21 ft. each, being simply stood up against the centre part and almost independent of it, forming together a building about 50 ft. long by 21 ft. wide. The age of the building and the good state of preservation of the greater portion of it determined its careful restoration and retention of its character as much as possible, viz. 'stud and plaster', as it is locally called: entirely built of oak, forming panels filled with 'wattle and dab', that is, round sticks laced or plaited together and plastered with clay and chopped straw.

The south end, being in a dangerous state, we pulled down and rebuilt partly with stone quarried on the estate and as far as possible with the old oak timbers, replacing the decayed ones by oak felled on the spot last fall. A wood-house and rainwater tank were added. The whole of the roof was stripped off, the rafters made good and the Horsham slabs were relaid, for which purpose many additional slabs had to be procured in place of those destroyed by wet and frost. Concrete foundations were put where possible to safely underpin the old walls. The ground floor was paved

1887 — The Moat House.

MOAT COTTAGE

throughout with hard red paving tiles on a bed of twelve inches of concrete.

Hollies.

Many handsome Hollies, gold, silver and fine green kinds, planted in December this year. They were got to shut out view of new kitchen garden from pleasure ground, and to save twenty years or more very finely grown specimens were sought. They were brought from Knaphill by road in our own and the nurseryman's wagons, which met and exchanged horses at Ockley: this to avoid injury to the finely grown plants from the several changes by rail. These plants had been frequently moved and came with good balls 2 to 4 feet across, the plants 6 to 12 feet high, many perfectly furnished

1887 — *Hollies.* and splendid in colour. They were planted in a row above the Scotch Firs on the north side of the pleasure ground. The immediate effect of the planting of the old specimens was excellent, well cutting off the view above and giving a beautiful line of trees. They are all the more precious from not being liable to suffer from severe winters, as exotic evergreens do. The soil we planted them in was very unlike that in which they grew in Surrey, which was a very free, peaty soil.*

Holly Hedge along Paddock. Planted in December, just within the new oak fence cutting off Lewis Mead and '5 Acres' from paddock, stout little bushes from Twickenham and Northampton. They are planted to form the best of screens and fence shelters for the east side of paddock: the intention being not to form a stiff clipped hedge, but a free line—at once shelter and good fence along the paddock, which would eventually cut it off from the cultivated field within. The little trees mulched at the root soon after planting to help them over the first summer and encourage free growth.

Oaks and other Trees in Pastures. Many of the Oaks in fields and around them had their boughs too low and in the way of the horses' heads. I mean trees which showed a 'browsing line', and yet in many cases were found to be a danger to the grazing horses and colts. In December, i.e. all the lower branches taken off so as to give free heading for any animal. In many cases these lower boughs were decayed or showing a very weakly growth, so that much of value to the trees was not lost. In any case, apart from the need of making the fields more safe for animals, the pastures are made more airy and pleasant by the 'boughing'.

---

* NOTE, 1894. *Many of these Hollies did badly, and many died: transplanting specimen Hollies is usually a mistake.*

A browsing line which is sufficient for oxen is not so for horses. 1887

Cardinal Willow.

Planted in Warren's Wood near stables, in January, 1888, about fifty young trees. *

---

* NOTE, 1894. *A source of interest and beauty at all seasons since planting.*

SUSSEX COW

1888

Medlars in Orchard. Planted group of Nottingham and Dutch Medlars on east side of orchard for beauty of foliage and flower more than the fruit. Could not get any on own roots in nurseries, so planted some vigorous and rough half-standards and a few standards all held together in a group on the outer side of orchard and coming between it and the Yews and drive.

Hazel in Woods. Planted many in the woods to make up gaps in underwood, Moat shaw, Stockards shaw, and replanted shaw at top of Pear-tree field wholly with Hazel, believing it to be the most useful underwood and less eaten by rabbits than Ash. Hazel not usually planted in this district, it coming naturally pretty freely. My plants mostly from Huntingdon—about 6,000 this year.

Farm or Park? The question of park or farm was several times raised in arranging the place. My friend, Mr. Marnock, the landscape gardener, thought the best thing to do was to make most of the fields into a park. After some consideration I came to think that the divisions of fields necessary for the working of the place as a farm was the best and prettiest way. The divisions and copses or 'shaws' have much use as shelter, and for a place of its size I feel sure its present plan as a manor is far better than any approach to the more pretentious park. At one time I thought of opening the whole of these high fields into one park-like expanse, but now see that the place would not be improved by such a change. There is no way so certain of making a place bald and limited as removing objects and dividing lines which give distance, and a park without great breadth and the dignity of great old trees is not beautiful.

Porch to Moat House. Commenced to make this from drawing by Alfred Parsons, April 24, of oak framing and rafters, with a view of making the old house prettier and more comfortable. Finished in May; put in two strong oak seats in old porch. 1888

Chimneys for Wood Fires.

The desire to have wood fires on the hearth being interfered with by the failure of the chimneys to draw led me to bring over a Frenchman to look at them. Our architects appear to have no principle to guide them in making fires of this class. Mr. Deevy had his own way with the fireplace and chimney in the new hall and failed to make that chimney draw to suit a large wood fire. Having noticed the good way in which fireplaces for burning wood in France acted, I found a 'fumiste', M. Pichon, in the Rue de Seine, Paris, and got him to come over to see if there were any means of getting the fireplaces to work well. He advised a totally different way of forming the fireplace to what is used with us, and the introduction of air from without. This air was to pass beneath the fire, and then around it in a flue-covered way, and then across to the front, discharging into the chimney inside the breast of the mantelpiece. There being plenty of wood in the place I wish to use it in preference to coal, preferring wood also for the beautiful fire it makes. Some harm has come from the fashion of making chimneys narrow, to suit coal and coke. Wood fires require a much larger chimney than coal ones do. The French 'fumiste' guaranteed that the chimneys would not smoke, if allowed to carry out his plans.

I began by experimenting on the Moat Cottage, and after some little difficulty, from the plan being unknown to our good English workmen, made smokeless in all weathers a very bad chimney. The French fireplace set in the Moat acted like a charm, and a room one was often driven out of by smoke became a very pleasant one. All the doors and windows being shut made no difference to its pretty hearth fire, as the good supply of air from the outside, flowing in beneath the hearth, and getting warmed all round the fire made it independent of the open door or windows. The practice of burning wood

1888 still being prevalent on the Continent has no doubt led to more effective fireplaces. It is much to be desired that in many wooded districts and well-planted estates in Britain we should burn the abundant wood often useless for the want of a market. For cooking and every purpose of heating it is pleasanter and better than coal.

Chimneys for Wood Fires.

Flower Gardening.

A garden should grow out of its own site, if we are to have the best of it. One should think of the spot and what can best be done in it, instead of following set models. My experience in forming my own flower garden may be of use to others. The things I had to consider were—

1. My favourite flowers.
2. Climate.
3. Soil.
4. Season of enjoyment.

The first question is: What do we care to grow? We can rarely succeed with numerous plants on the same spot of ground: it is against nature and experience that we should. Botanical and other gardens that attempt large collections are only kept up by ceaseless additions. And even the growth of a few groups of favourite plants must often depend on the soil and situation. My soil was cool, the climate cool, and the situation high and somewhat late for a southern county. The season of greatest enjoyment is the summer and early autumn. Spring flowers were planted in the grass to take their own way. There was to be no specially set out 'Spring garden'.

The old house prevented me from indulging in any extravagance of colour or mere collections of flowers. The simple form and the colour of the house were so beautiful that all had to be done in relation to them. Looking at all the circumstances I determined, while not neglecting the great variety of hardy things, to devote the garden mainly to a few groups of flowers, Roses, Carnations, Pansies, and Starworts, and in each case to special and favourite classes of these plants.

Tea Roses.

I wished to gratify my own taste for these things, and at

the same time do something to try and make them better 1888
known. I thought that these most beautiful kinds of Roses *Tea Roses.*
had been much neglected, and resolved to grow them largely and in the open air. The use of the misleading word 'Hybrid Perpetual' seems to have cast dust in the eyes of Rose growers, who, while paying much attention to the red Roses, seem to

WEST GARDEN, BEFORE THROWING IT INTO BEDS

have hitherto almost wholly disregarded the supremely beautiful Tea and China Roses as flower garden plants. They are grown as pot plants frequently, but the most beautiful Roses we have deserve the best place in the flower garden, where the soil and climate suit them. A large collection of Tea Roses was planted, hopeful of their hardiness and beauty. Of some seventy sorts planted out in the first two years none have been

1888 lost from cold. Roses are generally placed in what is called a 'Rosery', which is often a miserable neglected garden, so

Tea Roses. I resolved to put them in the flower garden in natural groups of from half a dozen to twenty to enjoy their long and delightful bloom. I never hoped to see anything so beautiful in a flower garden as those Tea Roses proved to be, though tried in some of the severest winters. In the Spring of 1886, after a three months' frost, men were skating on thick ice on the lake on St. Patrick's Day. The two following Springs were very severe, that of 1888 being followed by a Summer worse than Winter in its dullness, cold, wind, and rain. Even in that year the Tea Roses were lovely, though the buds rotted with cold and often could not open. Some of the kinds flowered four months in the fine Summer of 1887—Marie Van Houtte and Anna Olivier for example. Thus, we were able to prove several things of importance in flower gardening: (1) The hardiness, in South England at least, of the most beautiful class of Roses; (2) Their surpassing beauty as flower garden plants in long given bloom, graceful form and foliage, loveliness and tenderness of colour, such as no other Roses show; (3) To add to the number of these flowers by growing a number of kinds and finding new and distinct ones not hitherto grown in our gardens or nurseries.

Carnations. The reason for growing many of these was to show the great value of the self-coloured kinds for the flower garden as compared with the Carnations as the florists grow them. We collected the best self-coloured sorts from all parts and grew them in the open air without protection of any kind. The result was very beautiful in the end, though the first year's plants were all destroyed by rabbits in the hard Winter of 1886. I then made rabbit-proof all the garden and ground round the house, and lost no more from these pests. In the Autumn of 1887 about 2,000 plants were set out in various positions about the house, good layers. Their effect from the

first was good, the healthy rosettes being very pretty in colour. 1888
The season of 1888, the dullest and most rainy we have had in England for many years, suited them so badly that the flower buds rotted away. Notwithstanding this, as the season improved towards its close there was a lovely show of varied

Carnations.

LITTLE SOUTH GARDEN IN CARNATION TIME

bloom that lasted till late in October. One kind, the Comtesse de Paris, flowered until mid-November, and gave us large buds and flowers then. A race with such qualities would be splendid for the English flower garden, far handsomer than the old cloves and the few kinds hitherto grown out-of-doors. But as they are now, they were proved to be here the most precious of late Summer and Autumn flowers. In bold groups among

1888 — *Carnations.* the Tea Roses, and intergrouped here and there with other things there was splendid colour, such as no bedding plants gave that dismal year. It need scarcely be said that in fragrance, colour of foliage, association and value for cutting, they were far before all, save the Rose at its best. Out of the two thousand plants not one perished from cold.

Starworts or Michaelmas Daisies. These were a lovely sight in Autumn, 1888. Some years ago they were never seen, save here and there in 'bundles' in botanic gardens. Like the Lilies, before the good plan of growing them amongst shrubs came in, the Starworts were 'lost'. As soon as the heavy ground work all round the house allowed of planting shrubs, I dotted a number of the best Starworts through the Rhododendrons near the house, hoping they would furnish the bare ground between and bloom (unstaked) in due season. The result was charming. There was a long bloom of Rhododendrons in the Spring, but it was not nearly as pretty as the graceful wreaths of the Starworts as they waved in autumnal winds. The bad effect of the staking and bundling was got rid of wholly, the plants being sheltered by the bushes and their flower showed above them here and there. The effect was picturesque and natural in the best sense. When the Evergreens get too big it will be easy to move the Starworts into freshly set beds of shrubs and show their beauty in other ways. The above is not the only good way to show these hardiest of northern flowers, but it is a charming one, and it lights up a garden with much loveliness of colour in the Autumn.

Tufted Pansies. The beauty of the simpler coloured Pansies, the blues, lilacs, whites, and yellows, is so superior to that of the ordinary bizarre, striped, and wiry kinds that they deserve a place with Roses and Carnations. They come into flower so early, and continue so long, they are always cool in effect, fragrant and beautiful, and their colours soft and good. They flower for a long time throughout the Spring and early Summer months.

The tufted kinds were used, being planted in Autumn between Roses and Carnations; their dense habit prevented them from robbing or overrunning other things, whilst they sheeted over the beds with pure and tender colour. In a wet, cold season like 1888, they gave many flowers when other plants were half dead with cold. In that year plants put in the previous Autumn blossomed through the Spring, Summer, and Autumn, little clouds of delightful colour. Like all the other flowers I cared about, they were in colonies and bold groups here and there running together, never in lines and never dotted about singly. The reason I prefer these Pansies to the ordinary English kinds and the showy and often ugly French race is because some of them remind us of one of the blue Violas waving their many flowers over a turf, as close as wall-moss on the Alps in early Summer.

1888 — *Tufted Pansies.*

I call these tufted Pansies (instead of Violas as they have been called) to avoid confusion by giving the Latin name of the whole genus to a section of varieties. They are simply Pansies more tufted, closer and hardier than the older and larger florists' Pansies. They have been crossed or are crosses of the Pansy, tricolour, with some of the Alpine Viola, like Cornutas. But the Pansy itself arose from an Alpine or mountain Viola. By the plans of grouping adopted the colour difficulty vanished. There was good colour in the beds without the geometric monotony of patterns. There was sufficient mass, and the things being good in themselves, a colour scheme was seldom considered. When it was studied by putting things which harmonized in the same bed, the plan was usually set aside by subsequent plantings and without apparent loss. The colour question only forces itself upon us when confined to the exclusive use of a few plants which produce showy and formal effects. In such cases it is important to make the display as inoffensive as possible. I got more than the strength of colour of 'bedding out' without a trace of its stupidity, varying and

1888 pretty pictures throughout the Summer and Autumn, and the beds being fully planted in Winter, colour interest even then. There was the good blue-grey rosette of many hardy Carnations and the ever lovely leaves of the Tea Roses about the walls and on the beds, growing above carpets of mossy Rockfoil and green and grey Stonecrops.

Tufted Pansies.

Beds and Groups and their planting.

The beds, whether in grass or gravel, were large and simple in form. It is impossible to do good work in the little beds so commonly seen. Instead of cutting up a large plot of grass into a lot of small beds it is better to make large simple beds. If the bed is to be well planted it is absolutely essential that it should be large to enable the adoption of the grouping system and the forming of bold and picturesque masses. The shape and disposition can be best arranged on the spot and according to the ground and the taste of the owner. One object was to prove a variety of things as to plants that could be used near the house, and with the best effect. It is only by the use of hardy plants that the most beautiful effects and garden pictures can be got. The fine old wall south of the house was covered with native rock flowers. The narrow border at its foot was planted with low growing plants that did not detract from the beauty of the wall. Any showy gardening here would not have been right. The wall was left as nature had adorned it, save adding here and there a Rose or Fig, and a few colonies of green or trailing plants on the border. These were mostly hardy and evergreen so that the border might harmonize in colour with the wall and house at all seasons. Plants carefully grouped in picturesque ways are far more beautiful and interesting than if set out in the common mixed way. Almost everything was planted in groups, varying the sizes and shapes, holding some apart and some 'together' by 'turfing' plants beneath and spaces of repose around, letting other groups merge one into another, suitable plants intermingling. This destroyed all set pattern to such an extent that it was impossible for the eye to take in the arrangement or contents of any single bed from one standpoint. While the arrangement of a geometrical garden is seen at a glance, it was here necessary to walk around each bed to acquaint oneself with its contents. The beds in the turf have a grass edging, but those in the gravel are edged with thin pieces of native sandstone. This forms a hard and durable edging which becomes clad with moss and lichens of varied tints. This simple edging is far superior to anything in the tile way. Many Alpine plants, and especially the Stonecrops, were found most useful in 'carpeting' the surface of the ground. In this way, while themselves beautiful, they add to the charm of other flowers. They were especially useful in holding groups 'together' as when the ground beneath a group of delicate Tea Roses is surfaced with dwarf and hardy plants. With no rocks in the place many of the vigorous Alpine flowers thus found a home. Many of them are pretty in colour, even in Winter. They are also very useful in helping one to get light and shade in the beds. So cool and refreshing are they at all times that they were found useful as a 'relief' by having a piece of ground merely surfaced with one or more kinds, and left to form a little lawn between the flowers.

1888

—

Beds and Groups and their planting.

Fenced with iron and wired and planted margin of the south side of upper pond, to give some covert for water birds. This, with the close fencing already done, completes a protected area around the house in all directions, cutting the garden off from the attacks of rabbits and hares, even when the water is frozen. The parts not wired are protected by a sunk fence. With care, Hollies and other things often killed by rabbits will be safe. This plantation was done with a view to destroy the sameness of the sides of the upper pond. In this way considerable difference is established between them.

Protection from Rabbits.

1888

The Eighty-six Acres.

Bought this estate, along side of and originally belonging to Gravetye,—long wished for by me to allow of the making of a good and easy road into the place. The ground being high and above the old house, I thought it essential to add it to the estate: it has high-lying fields with good soil, though long neglected. These being near the farm-yard and stable will be more profitable to cultivate than the distant and hilly fields and the heavy ground. With these high fields, where sheep will thrive, better farming will be possible in future. It has fine views over the country near a Larch wood, which will form a picturesque fringe to the drive and will allow of good views of the North Downs.

Tile Barn.

At one time it was a question of removing this old barn and yard, but those whose taste I trusted best said, leave it, and now it has turned out very useful. This year we began to repair it, roof and old wooden sides; spouted it all round instead of letting the water fall off the roofs, and made a bog as before: raised the inside with sandstone, dug in the field near, also the wet yards outside and around: fitted the whole as a sheep yard and sheds (for lambing): tore away the old barn floor and fitted it neatly with troughs for sheep and lambs: drained it all round and cleaned out water pit at back and altered doors. The barn is a fine and very old shed, with massive oak beams, and well worth preserving.

Rights of Way.

When I went to Gravetye every field in the place and even the garden round the house had rights of way, or reputed rights of way! People passed under the windows of the house, and any garden made round the house was crossed by gangs of men, with dogs betimes. New and good roads and railways have made many footways of no real use to the public. A few generations ago there were no good roads in many districts that are now well supplied with them. In many places there are supposed rights of way in every field. If these were allowed no fair and full use of the ground could be made for choice stock or crops. Frivolous claims of this sort are the cause of wrong and annoyance. Ways or avenues made for the use of an estate should not become public paths through lapse of time merely because they have been trespassed upon. Save the commons and the grassy roadsides by all means, also every way necessary for public convenience, or even for seeing the charm of picturesque or beautiful districts; but to interfere with private rights, with stock, with delicate animals in young, with game, is not just to those who have to bear the burthens of the land. The hope of our farmers is in good races of animals, which are known and sought throughout the United States and our own Colonies, and also in many parts of Europe. The life of these animals is, for the most part, passed free in the fields, unlike the tying-up system so much in use abroad. Can the farmer do these justice when the poacher and idler and their dogs have the right to be on every footpath at any hour of the day or night? Often gates are left open so that stock pass into crops which are nearly ripe, perhaps for harvest, and not only is damage done to these, but frequently the animals themselves suffer from gorging themselves with improper food. The dog gets a free run on these bypaths and often does cruel work among pregnant ewes and in-calf cows. Any public authority having a right to keep necessary paths open should have the right to stop frivolous claims of rights of way and to see that justice is done all round. Now, if the large farm through which these paths go is less valuable through these paths, there should be power to close them. The age of paths should be no proof of their necessity, as changes in manufacture make paths and roads once essential now useless. The woods of many estates are scored by old and forgotten roads, now densely grown with trees and brambles. To prevent the closure of a footpath claimed as a right of way it should be proved that it was required by the inhabitants of the neighbourhood. In such instances as where a good road exists on two sides of

1888

Rights of Way.

1888

Rights of Way.

a field, a short cut across the field should not be allowed as a right of way.

In the case of Gravetye people came into the garden to offer things we did not want, and there was less privacy than in a villa garden near town. They went as far as cutting down valuable shrubs. I felt that some of these claims were not sound, that the people had simply been allowed to come so near the house from carelessness on the part of previous owners, and from old ways to the house having become rights of way. I determined to fight for entirely freeing the space around the house of such annoyance, and in the end succeeded in clearing six or seven acres of any such intrusion. I felt that if I did not do so the place would not be an enjoyment to me, or to others who might follow, though I had no objection to people coming to the place in one or more necessary ways, that they might enjoy its beauty. Generally the people did not object to what was done, and thought it fair, but one man in the village, who never used the paths himself, sent some men to force their way through the garden and damage things generally, and they did a good deal of harm. They were seen by the keeper and bailiff, so I proceeded against them:—

History of the Case.

WILLIAM ROBINSON *against* LEPPARD WALTER WREN, CHARLES TAYLOR, EDWARD DAY, ALBERT GOLDING, EDWARD BILLINGS, JOHN LANGRIDGE, and JAMES POLLARD.

In the year 1885 there was upon the Gravetye Estate in the Parish of West Hoathley, a footpath, the position of which is shown upon a plan then made by the red line between the letters A & B, B & C, and B & D. The said footpath was made and used solely for the convenience of the owner and occupiers of the mansion at Gravetye, and was not a public way or a way over which there was any right of passage, except with the consent of the owner of the said estate. In the year 1886 I discontinued the use of the whole of the said footpath, and laid out a portion of the estate, shown on the plan by the colour green, as a pleasure garden, and planted the same with valuable shrubs and ornamental plants.

The above-named people alleged untruly that there was a public right of way over the said footpath, and that such right existed over the site thereof. A few days before the 9th June, 1888, in pretended exercise of such alleged right they passed over the alleged site or thereabouts, and so over the Warren Field shown on the said plan in the course or route shown by the dotted blue line on the plan and destroyed some of the shrubs and ornamental plants and committed other injuries to my property, notwithstanding the remonstrances of my men.

1888 — *History of the Case.*

The above people having threatened to repeat such acts in support of their alleged right of way, I instructed my Solicitors, Messrs. Cripps, of Tunbridge Wells, to take proceedings to restrain them from their threatened wrongful acts, and on the 9th June, 1888, a Writ was issued in the Chancery Division of the High Court of Justice claiming a declaration that the said footpath, the site whereof is shown in the said plan, was not a public footpath, or a footpath over which any person had the right to pass without the consent of the owner of the said Gravetye Estate, and an injunction to restrain the above-named people from trespassing on the site of the said footpath, or on any other part of the said Estate in pretended exercise of any such alleged right of way or otherwise. Such Writ, together with the notice of motion to the Court for an order in the terms of the prayer of such Writ, were duly served upon the above-named people.

The defendants did not enter any appearance to the Writ, or put in any defence, and on the 28th day of July, 1888, Mr. Justice North, sitting in Court upon Motion for judgement, adjudged that the footpath situate in the Parish of West Hoathley in the County of Sussex, shown in the plan annexed to the statement of claim by the dotted blue line on the said plan, was not a footpath over which the persons above named had a right to pass without the consent of the Owner of Gravetye Estate. And further ordered that the above-named persons and every of them be perpetually restrained from trespassing on the site of the said footpath or on any other part of the said Estate in pretended exercise of any alleged right of way or otherwise. And he further ordered that the above-named persons should pay my costs of the action, to be taxed. A copy of the judgement was served upon each of the defendants.

SCARP OF ROCKS BROUGHT INTO VIEW BY REMOVAL OF UNDERWOOD COPSE, MILL PLACE

1889

Mill Place Estate. Completed the purchase of this on the 16th April, as land likely to make the Gravetye Estate more interesting, and the whole more valuable for farming and sport. Much of the land is bottom land, and far better than any part of the original Gravetye Manor Estate, the woods of the two places running together; the whole now forms a compact estate with the woods at once pretty and extensive, and lying very well for game, and the fields for stock.

Mill Place, like Gravetye, was formerly in Elizabethan times the seat of ironworks, and there are now many tons of slag lying in heaps near the farm-house. Cannon and shot were cast here, and the men still find many cannon balls on a bank near the bridge, at a spot where it is said that cannon were tested. The Mill Place ironworks were working up to the year 1653, but ceased then or very soon afterwards.

The timber on the estate was very poorly managed, and much timber, too old, and decayed in many cases, covered the old house and pretty banks near with gloom. Many useless fences and remnants of fields divided by the railway had to be removed and so I cut down a good deal of timber with a view to many changes in the landscape in the Autumn. Among the timber was a great quantity of Spruce, which never seemed to me to do well in the upland of the district, so I cut down every tree of Spruce: also many Wych Elm which did not seem to have its usual fine character or to grow well here. Larch too ill-placed and grown to fullest age was for the most part cut down.

Grafting Trees. In the Spring of this year many attempts to get standard Apples on their own roots began to be successful, and I had the pleasure of seeing a number of Ribstones, Cox's, Stone's,

GROUP OF ASH, MILL PLACE

and others, rooting independently of all stocks, from layers and trees thrown down on their sides. There is not a greater nuisance to the practice of gardening than grafting. It exists mostly for the convenience of the nurserymen and is not only needless but harmful in most cases. Grafting enables the nurseryman to supply stock of a certain variety more quickly than he otherwise could do. It is harmful to the tree itself and the trouble of the continual removal of suckers is great. At first layers of some shrubs are slow, but once started they

1889

Grafting Trees.

MILL PLACE FARM HOUSE

make up in vigour. Then they are not confused, as the grafted shrub is, in struggling with an enemy below them, like the grafted Rhododendrons, with the multitudes of shoots that come up round it—a struggle which ends in death. Not only are distinct species put on each other, but the most incongruous things, as the Medlar upon the Hawthorn, and the Pyracantha upon the Quince. I bought a number of Medlars for ornament and was surprised to find thorns coming up from their roots. I had prickly Pyracantha for fencing. Towards the end of Summer Quince shoots came up from each! The Quince being vigorous in moist soil, the Pyracantha, which is only an evergreen trailing shrub, has a poor chance. The beautiful

1889

—

Grafting Trees.

rose Acacia we never see in its natural condition. It is often stunted and distorted, and in a strong wind the head is blown off. Layering would, no doubt, give more trouble to the nurserymen than the expeditious process of grafting, but then it is their business to take trouble. Roses are often planted to die. I bought a bunch of Catherine Mermet and planted them in a rich border, and after a time found they were on the Manetti stock. After three years the plants had scarcely grown at all.

The grafting of ornamental trees is an evil without excuse, and without a redeeming point. People generally plant in ones, and the system of planting a shrubbery is such that deaths are not noticed, or, if noticed, they are frequently put down to canker, cold climate, soil, and so on. Many of the beautiful flowering trees and shrubs introduced into England are lost through this practice of grafting. Nurserymen are accustomed to have a lot of stocks for general fruit-growing uses, and they adapt them to ornamental trees with little regard to consequences, beyond getting rid of their stock. I made a large plantation of beautiful shrubs, and as I plant in groups each thing by itself, quite distinct, I saw the effects of grafting more than is often afforded in private places. The plants were bought in some of the best nurseries in England, and in almost every case death or injury resulted. I find it difficult or impossible to get things on their own roots. Shrubs that, like the dwarf Almond, one would naturally like to secure in their dwarf state, are grafted on such strong-growing things that the union seems a hopeless one. It is not only that we lose the things by disease or canker, but if the shrubs do grow on these strong-growing stocks, we are prevented from seeing them in their natural habit and size.

The New Upper Drive to House.

Finished May 11, 1889, after very heavy labour for horses and men for eight months: felling trees, filling hollows, removing many thousand loads of earth to get a good line of view and an easy grade, closing old and useless roads, so as to make the approach simple as well as beautiful and easy. Line brought below Bailiff's cottage to get easier grade, and much ground taken from Warren's Wood, the bank being raised here as much as twelve feet. After the heavier work of levelling and getting the grade was done the whole line, as far as the Farm, was dug out to sixteen inches deep and one foot of sandstone put in. This sandstone was dug on the place in one quarry beside the drive and in another in Warren's Wood near. Six inches of flint, and two inches of hoggan, both from Croydon, were put on and carefully set and rolled. Twenty-two new gullies, grates, and drain pipes were put along the road, and many along the side entrances to same, such as up-hill from the stable towards the Moat, which hill was drained to try and keep away the frequent rushes of water here. In addition to thoroughly draining the road we cut an open 'dick' a little above the road to stop any rush of water down the high fields above. The road from Kingscote to the present main entrance of Gravetye broke through and we had to repair it. The various quarries and sand-holes were filled, with much labour, and the fields on each side enlarged and simplified so as to fall prettily into the landscape. Owing to much ground of this lying above the drive, many precautions with drains and ditches were taken so as to keep the rush of water off the drive. The open ground along the drive was planted with dwarf or low native shrubs, Heather, Gorse, Brier, Bramble, Broom, Sloe, Cherry, Plum, Dogwood, Spindle tree, Barberry. In the woods in parts taller things were planted to fill gaps, Ozier, Mountain Ash, White Beam, Chestnut Ash. The drive is to be fenced with our own split-Oak posts and rails, so that there may be no need for iron fencing here. Lower side of fencing completed by May, and very handsome and well made by Brazier, an excellent type of the Sussex woodman. The views resulting from cutting away a few

1889 — *The New Upper Drive to House.*

1889 — *The New Upper Drive to House.* small plantations are very wide and fine. A place that never had a good road for some hundreds of years has now as good a one as I have ever seen. The feeling of 'air' and the views give one quite a different impression to what the old roads through the hollows and the woods gave.

Furze, Broom, Acacia. Sowed seeds of these on great banks by railway at Mill Place, scattered them with no preparation or cover of any kind. June 9.

Lower Lake, Wall of. Saw defects in brick-work; here and there were serious leaks below, which kept the lake below its natural level in Summer. The whole thing was in such a state that it was in danger of falling down. This was owing to the filling up behind the stone and brick-work rubbish which was washed out by the water working through here and there. Pulled down wall at south end and re-built same, 16 feet long, 12 feet deep, and 4 feet 6 inches thick, well puddled from south corner to overflow in centre on lake side. Where the water flowed below the overflow was levelled, and filled in with rough concrete, the remainder of the large Yorkshire sandstones from the Manor House put on top and finished off with brick on edge laid in cement, and the two side walls were built part dry and part with cement.

Foxgloves. Sown in the woods and on broken ground near their fringes in hope of establishing the pretty coloured forms among the wild one.

Columbines. Sowed some seed by new drive: would prefer seed of the wild English form: also in other parts by new lawn over sunk fence.

South Down Sheep. Purchased choice flock at Aldwick Farm, near Bognor, bred from best flocks by Mr. Henry Ayrton of that place, at prices from £3 15*s*. to £4 12*s*. per head. Bought ninety ewes in all. August 7 and 31.

Poplars at Tile Barn. Planted a group here to screen the sheds a little from the house. There was one Poplar tree already in the spot about thirty feet high. It suggested a few more of the same kind there: they were tied to some Larch trees that did poorly there, so we cut off the tops of the Larch and planted the Poplars against the trees, using these as stakes. 1889 — *Poplars at Tile Barn.*

Commenced to form hereabouts a series of bold beds to screen the rather steep and awkward banks, and at the same time to give a home to many fine things that would not be very well placed in the flower garden:—Bay, Pampas Grass, *Shrubs above Play-ground and on Slope over Flower Garden.*

GROUP OF SCOTCH FIR, MILL PLACE, AFTER CLEARING AWAY OTHER TREES

Barberry, Venetian Sumach, and many choice flowering shrubs planted in November. Planting of tall flowers and Lilies between to go on all the Winter.

After much trouble with fences, seem to be arriving at 'firm ground', making fences this Autumn of a bank with very low sided 'dicks', forming a good bank to keep sheep and cattle off, and setting Sloe and Ivy along the front of the banks, as they are made up, so as to tie the bank together. Should never again make an iron or mechanical fence, if it were not for the rabbits. Propose to use Sweetbrier and Dog Rose as well as *Live Fences Best.*

F

1889 — *Live Fences Best.* Quick, Sloe, Holly, and Cherry Plum. There is no kind of fencing which is half so good as the live fence, or so enduring or so beautiful. Mechanical fences give no shelter and are objectionable in other ways. Our own Quick is excellent, and they have lately discovered a number of Quicks in America with fierce thorns, which some day will make excellent fencing plants.

Rocks in Mill Place. Found group of rocks here obscured in many parts by trees, and hidden by a stiff Chestnut copse; removed it and kept the Holly, Yew, and Scotch native Evergreens of the place only, intending to let the pasture run up to the rocks.

Meadows in Mill Place. Began to get better effects here by removing or changing the direction of fences that hid the views of the beautiful meadows without sacrificing the use of the fields in any way. Did this and got fine long reaches of meadows in several directions.

A New Aspect of Garden Beauty. Gardeners as yet seldom look at general effects or at the whole of things. The flowers are so dear to them that the garden, as a picture, is left to chance, and hence so much ugliness in gardens, for those at least who look to the robe as more than the buttons. Some years ago Starworts were rarely seen, save in bundles in botanic gardens. Since the hardy flower revival they are more frequent in collections, but as yet have no important place in gardens generally. As soon as the heavy ground work round the house allowed of planting belts and groups of flowering shrubs, I dotted a few of the best Starworts through these, to furnish the bare ground between, and flower in due season. The result was charming. If some kinds were dwarfer than the shrubs it did not lessen their value, but on the other hand, increased it, as it helped to give light and shade, and destroyed the common way of setting plants to a face, as if one were laying bricks. When the American plants close together it will be easy to move the Starworts into fresh beds, and to show their beauty in other ways. This is not the only way of growing these hardiest of Northern flowers, but it is a charming one and it lights up the garden with a loveliness of refined colour not before seen. Being so well pleased with the results in a few small plantings, in the early days of this year I had from various friends and from botanic gardens probably all the Starworts in these Islands, to make trial of their merits. They came with and without names, tall and short, good and bad. Like many other flowers they vary a great deal, and there are many names, so that it was well to find out the best of all those now to be had, and to give the chosen good and simple names. The effect of the greater number of plants put out in the Spring of the present year was very pretty. We had lovely colours, showing well through large masses of shrubs, and seen in the distance, while the colours were well held together—i.e. naturally massed flowers of one kind. The colour was both telling and refined—absolutely different, it need scarcely be said, from the pattern stuff of the ordinary flower garden.*

1889

A New Aspect of Garden Beauty.

December 17 and 18. Planted Sweetbrier covert, i.e. the upper flanks of the lower lake, with dwarf Spring flowering and other shrubs—Sweetbrier, Sloe, Furze, Siberian Dogwood, and Barberry. The aim was to afford shelter to birds about the lake: also to plant in a pretty way steep banks of little use for grazing, and connect the upper and lower lakes with a belt of low covert, which would help the birds and be no loss to the home landscape view. Planted at the same time a sort of trial fence, Hollies, at west end, running into Barberry, Sloe, and Sweetbrier, and at last Cherry and Plum, as a test of various fencing plants under the same conditions.

Sweet-brier Covert.

---

* NOTE. *The names of the best kinds are recorded in the book, but as they are changed from year to year and there are other kinds and hybrids being raised, there is no need to enumerate them.*

1889 — *Sweet-brier Covert.* On the opposite shore at the same time planted common Rhododendrons and other dwarf Evergreens, so as to clothe the ground and keep the view from the field in front of the house quite clear. Fenced the off side with split-Oak post and rail, and below it a good line of strong gapping Quicks.

The Japanese Arbor Vitae. This beautiful tree (*Thujopsis dolabrata*) is invaluable for pleasure grounds and woods. Even when Conifers do well (many of them do not) they are entirely unfitted for this country. When the gardener digs deep holes for them, and surrounds them with four feet or five feet of rich moist loam, they thrive very well for a time; but once they get into the natural soil of the district and have to submit to the actual conditions of our climate, then it is seen that many of them are unfitted for it. Many of them too are tender and get killed by frosts. But this tree is as hardy as our own Holly. When so many trees were killed in France in the terrible winter after the war this tree was uninjured. Moreover, it is more beautiful than many of the hardy Conifers: the foliage and form are so distinct and graceful. Those who can get good plants of it will do well to form a bold group of this tree. When young it is somewhat shrub-like, and the danger which so often occurs in our gardens of overcrowding is present. This tree, when young at least, keeps its branches near the ground, and roots very freely where it touches the earth, so that in this case one has some chance of escaping the grafted plant, far too common among coniferous trees. It is a beautiful tree in its native land, and when young is gracefully furnished to the height of fifteen feet or twenty feet until it becomes a tree, and like many other Conifers loses its lower branches. To the ordinary gardener this seems a loss, but it is natural that the trees should lose their branches. The great Pines, whether Scotch Firs or Japanese Arbor Vitae, are never so picturesque as when they have lost their lower branches.

Planted 150 of the American hardy *Rhododendron catawbiense*, to watch how it grew on the same bank with the pontic Rhododendron. 1889

Covert at Top of Large Lake.

The White Willow.

Planted many this year. Among the trees of the northern world I think there is nothing more beautiful than our native White or Grey Willow. People who ransack the world for trees commonly do not take so much notice of their own; but none that we can ever bring home from the farthest earth will be more beautiful than this for our landscape, or more useful in our woods. Those who know the Eastern Counties will have seen many beautiful examples of it in Essex and in the Eastern marsh country. Occasionally we find large specimens of this tree in some or one of its varieties, in rich bottom lands and in parks where it has room to grow; but, after its native swamps, it is most beautiful, perhaps, in groups and colonies in woods, where the effect of the leaves when tossed by the wind is as pretty as that of the Olive tree in the South. It is well worth growing for its beauty, but it has uses as well as beauty, making excellent cricket bats, and in other ways; while, as underwood, it is readily bought in a younger state than almost any other. While the ordinary underwoods cannot well be cut before nine years' growth, the White Willow makes very good growth at seven. The simple way is to take some bundles of stout faggots and drive the sticks nearly half-way into the earth at about the distance we wish them to stand: they then grow rapidly—in fact very much better than young trees from a nursery. A curious thing about Willows is that they seem to make better plants by driving in branches from trees which have been chopped down than by planting young trees. Another charm of the White Willow is its delicate and grateful odour. The pale green-yellow catkins are scarcely visible among the green in Spring until a delicate and most welcome scent lets one know that the trees are near. It does not last long, but in all the charms

1889 — *The White Willow.* of the early year there is scarcely anything more welcome to us than that of the odour of the White Willow. Probably few people have noticed the flowers, as they are so much like the budding leaves in colour. Gilpin is very shallow about it, and in his description of the effect of the tree in any rural scenery he misses its beauty altogether on the marshes and in picturesque groups elsewhere. It is a tree of wide distribution in Europe: trees seventy or eighty feet high are not uncommon.

1890

The True Laurel.

It is unfortunate that the common Cherry Laurel, a nuisance in many places, should have gradually taken the name of Laurel, to which it is not entitled. The Cherry Laurel is a fine Evergreen when well grown and grouped, but it is a rampant grower, and so destructive to other things that it has become a source of monotony and of loss in many places. When with other shrubs it kills them, as its growth is often too rank. People then get relief by cutting it down. The effect is not pretty in either case. I know nothing more dreary than vast areas of this shrub cut with the shears to ugly slopes. In very hard winters it is killed level with the ground. In some places the beauty of the planting is destroyed by this plant, which is one of the robbers of the soil. It is equally unfit for little gardens, where there is not much room for gross feeders.

The true Laurel (*Laurus nobilis*) may be found frequently as single specimens; but few people think it worth their while to plant it in masses and groups. There is nothing more delightful than a colony of this most fragrant of Evergreens. It is interesting in every way, not only as a garden plant, but for associations of the highest order as well as for its beauty. There are several varieties of it in addition to the common form. It requires some care in transplanting, otherwise the plants are a long time rooting well, and sometimes they die. It is a well-known Eastern shrub, and in the classic age was the emblem of victory, and also of clemency. The victorious generals were crowned with it in their triumphal processions; every common soldier carried a sprig of it in his hand, and even the dispatches announcing a victory were wrapped up in and ornamented with leaves of Bay.

1890 — *Birch Farm.* Knocked away some of the cross fences here to simplify the fields that had been cut up by the railway. Refenced the garden and simplified its outline. Grubbed and runted a good deal of ground about the house to make all the fields useful, clean, and fenced in the most simple way we could manage, at once for the good of the fields and convenience of working, also for the beauty of the landscape.

Larch Wood planted. In February planted Larch two to three feet high (27,000) in Marl Pit Field, near Birch Farm, 6a. 3r. 13p. The field was an awkward one to get at and not productive, so I thought the best way was to plant it with Larch. It contained a very large 'pit hole' from which marl was dug in past times, and by planting this at the same time with Scotch Fir we simplified the whole space about here. The Scotch Fir will afford some evergreen covert in the great 'pit hole' as it grows up, and probably grow better than the poor Oak we cut out. About 2,000 Scotch Firs were set at the same time in and about the 'pit hole'—making in all nearly 40,000 trees at Birch.

Rocks Field in Mill Place. After the grubbing and clearing needed to show the rocks, fenced the upper or north side of this field, using Holly, Sloe, and Quick: also the west side where a few native Hollies suggested planting more there to make the fence good. The effect of the fine wreath of rocks over the level meadows by the stream is good.

The Mount Field near Railway. Grubbed a stiff old fence dividing two fields here, and threw all into one as simpler in effect and easier to cultivate. It is curious to note the great differences of level between two old fields originally, doubtless, one surface. These differences are caused by the plough working from the fence for ages, especially where on one side there is a downward pitch of the ground.

1890

Wood near Birch

Corsican Pine.

Began to plant a field of this, having seen it do well in France and elsewhere on poor soils. This field not being a good one and almost surrounded by woods, I thought it would be better as an evergreen wood, the more so as all the woods around are summer-leafing trees at present, so that it may give a welcome shelter to the birds in winter. Began to plant rather late. Did not order the trees till March. Name of the field, Old Mead: a little over seven acres in area. Propose to plant in all about 15,000 trees.

Native Shrubs for Covert.

It is difficult to get away from the usual nursery selection of trees and shrubs, as any one who tries to plant any of our own beautiful native shrubs soon finds. The dwarf Furze (*Ulex nanus*) is common on many heaths in the south of England and elsewhere, and also towards the west. There are many introduced shrubs less valuable than this, but I could not get it in any nursery. There is a Welsh form of it called *Ulex gallicus*, and the two are distinct in time of flowering and habit from the common Furze. Another bush which is common enough in our copses is the Water Elder (*Viburnum Opulus*), the single or parent form of the well-known Snowball tree of the shrubberies, and a more beautiful tree than the double form. As it grows in the underwoods in Sussex and also crowds by streams, we get its name of Water Elder. Mr. Anthony Waterer told me that last year he got an order for one hundred young plants of this, and after searching throughout the trade, failed to get them! It is often best, in planting a large place, to use native shrubs. For instance, by rough drives and by drives away from the house it is inartistic to introduce exotic plants if native ones can be obtained, and very often Rhododendrons even look out of place where a colony of the red Dogwood, or of Furze or Broom, would be very pretty. Among those I planted this year and last are:—

1890

Red Dogwood. Is a native of Britain, and a pretty shrub for massing and grouping by drives, and indeed, wherever a bush is required. It is not the same as the Siberian Dogwood.

Sloe. Of this, several thousands. Its pearly little flower-buds in Spring are almost as precious as the Hawthorn blossom, and a colony of flowering Sloes is very telling in the distance, as well as very pretty at hand. It was largely used in fences this year.

Cut-leaved Brambles. An excellent plant by drives, or in any position we want to make a little difficult for people to get about in. It is not easy to get this in nurseries by the thousand, but it can be had here and there. Sometimes the question of protecting lakes and other places from interlopers may be settled by a good mixture of Sweetbrier, Sloe, and Furze. Sweetbrier is a splendidly armed bush in soil where it does well.

Broom. The Broom will often be found extremely useful for covering a poor bank, and is very effective. If bought strong in nurseries, it transplants badly, and often it does much better from small seedlings picked up in one's own woods. This is also true of the Furze, which is a most valuable plant for this sort of work. Planted several thousands this year in places where bushes only are desired.

Double Furze. Several hundreds between the ponds. The double Furze might often be used to great advantage, but it transplants so badly that it must be bought in pots, which prevents its use on a large scale.

Sea Lyme Grass. Some scores near new drive. A bold, vigorous plant that takes hold of any poor ground, and makes good covert. The smallest bit roots, and it may be readily divided. The effect is good either in winter or summer. I do not know any other plant that will so prettily cover some deplorably arid or sharp ugly bank as this.

The Common Raspberry. Some covert plants are vaunted as food for game, but without reason; but wild Raspberries and Blackberries will always be the food of birds. The effect of the fruit is charming by a stream or about an old bridge. 1890

The Common Bramble. Everywhere in coverts, but few act upon the fact that there are a great many different Brambles that are natives of Britain. They differ considerably in their variety, qualities, and other ways, and some of them would make excellent covert plants. They are sometimes local, occurring here and there, and not cultivated in nurseries.

Sweet-brier. Of this charming Wild Rose over 5,000 have been planted up to this date in groups, fences, and coverts. Whether it will thrive as well as on chalky hills and warm light soils remains to be seen. Also many of the Dog Rose more frequent naturally in coverts.

Honey-suckles. The native kinds by new drive, and to cover the ground—not to climb as they do in the woods. The plants are easily picked up in the woods.

The Dyer's Weed (Genista). Planted this by new drive, too common in some of the pastures, and easily gathered in them for planting.

Tamarix. A few by drive.

Spindle Tree. By drive.

Sea Buckthorn. By drive.

Sweet Gale. Moist hollow in pleasure ground.

Good Fences. I have made several miles of fences on Gravetye Manor and Mill Place estates during the past winter. The art of good fencing must be understood in any country place if there is to be any peace for the owner and his men, or safety for his stock and crops. I have made many experiments in fencing and seem at last to be arriving at firm ground, and that is, that no fence is so good as a bank and a dyke with prickly shrubs on top of it. A good bank is certainly half the battle, and in all fences made of late we began by making a good bank the size to suit the nature of the ground and situation, except where the ground prevented us. The native shrubs cover all awkward banks and angles, and grow all the better

1890

Good Fences.

for being on them; the advantage of getting free of nails and mechanism in every shape is great. It is a clear gain to be rid of them for ever, for the sake of the stock, if for no other reason, as many accidents have happened from pieces of iron and old nails being either trod on or swallowed by cattle. Iron fencing, as now often used, is robbing the country of its beauty, and cattle and crops of shelter. It is not well that the blight of iron should come near the house. People are often mistaken as to the endurance of iron fencing, but I have seen an iron deer fence moulder away like rotten sticks. Unless it is very strong it soon becomes only fit to scrap, and nothing is worse than broken iron fencing. The cost in the first place, if well done, is very great, and its ugliness is a serious objection. Properly made live fences are far better than the most costly iron fence. Iron wire fences, which we made a good trial of, are perhaps the worst of all. They are no protection against sheep, and horses and heavy stock knock them about. They begin to rust in a couple of years, and the iron not being very thick is soon eaten through. Horses also get their legs entangled in the wire. Rent Oak (of which we have made the finest examples of fences I have seen) forms a grand fence and most valuable in certain positions, but is too expensive for general use. A living fence of some of the hedge plants must be planted in line with it to make it effectual; such a fence is good against all kinds of stock, and with care will last many years. We used it where a strong barrier was essential, as along the new drive and where I would not put an iron fence.

Holly as Shelter.

A fine fence is unclipped Holly which is almost as good as a shed in sheltering stock from cold winds and driving storms. Where valuable animals are kept it is often well to conceal them from each other as well as to separate them. For this purpose nothing equals a good bank with a Holly hedge on top of it. We used Holly by itself and also intermixed with Sloe and other plants. Holly is sometimes slow in starting, but once it begins it is a quick grower in good soil. Vigorous seedling Hollies are used for the fences, preferring those that are about one foot and a half high. In making shelter fences and belts I would not hesitate, however, to plant bushes as high as five or six feet, if I had them and could plant them at the best time, i.e. May. Those who plant much Holly in fences should raise a stock of their own as it is often very difficult to get good plants in nurseries. Some say Holly is not eaten by stock, but ours was eaten greedily by young horses, and fences to be safe when young should be on a bank or otherwise fenced to keep young horses off. Sloes put in two years ago are now a mass of needles, and young horses that eat almost everything else within reach leave these alone. The Sloe seems to grow quite as well as Quick, and no plants could be more sturdy or better armed. Sweetbrier has been largely used to mix with other fence plants as it is fiercely armed with prickles and is so valuable apart from its other charms. Hornbeam forms a good fence when mixed with Quick or other shrubs, and is perhaps the next best plant to the Holly for forming sheltering hedges, its dry brown leaves keeping on all the winter. We use it largely mixed with Quick and Sloe. Quick we used many thousands of, both for gapping and forming new fences. For a farm fence it is not to be beaten, except in the point of shelter in winter. We made some very good strong fences by digging strong Quick out of the underwoods and putting them on a bank rather close together, with an odd Holly bush here and there among them. If these get through the summer safely we can hardly hope for a better fence. In large places it would be an excellent plan to have a stock of strong Quicks coming on, with which fences could be made so strong that no stake fence would be necessary to protect them. Ivy was used here and there to 'tie the banks together' and run through the bushes. It is often charming in

1890

Holly as Shelter.

1890 — *Holly as Shelter.* a Whitethorn hedge, giving a welcome green and shelter, while the Thorn keeps it down and prevents it from getting too strong. Furze we also used, but as it is not quite hardy in severe winters, nor very long lived, we do not trust much to it. The various new lines of fence plants, whether on banks or not, were generally protected by the usual 'stake and Heather' fence of the country. I tried hard to get something less fragile than this and failed, but when a good live fence is behind it, all is well. In one case we made a strong fence, without any protection from the stake and Heather, by having a good bank and putting stout bushes close together in the line, to prevent stock brushing through, and we ran a Larch pole along just below the level of the hedge. This made me think that it would be best to have a piece of ground with stout and well-grown fencing plants coming on and ready to make a strong fence at once, so saving the great trouble of 'stake and Heather' fencing.

Naming Stock. Christened some of our own earliest born Sussex, and also bought heifers, by names mostly of flowers and places, such as Primrose, Meadow-sweet, Violet, Marsh-Marigold, Jonquil, Daffodil, Pansy, where a fitting name was not suggested by the origin of the cattle.

Planting in Season 1889–1890. Apart from Roses, garden flowers, and shrubs, about 120,000 trees, underwood, fencing, and covert plants were set out. Larch, Quick, Sloe, Hornbeam, Cherry, Plum, Sweetbrier, Willows, Scotch Fir, Corsican Fir, Calabrian Fir, Black Spruce, Blue Spruce, Dogwood (Siberian), Alder, Holly, Clematis, Furze, Rhododendron, Ash, Sycamore, Honeysuckle, Ivy, Heather, Dyer's Weed, Spurge Laurel, Broom, Chestnut, Hemlock Spruce, Red Cedar, and a variety of American trees and shrubs.

Green Chisel Pear. Planted three of these among the Scotch Firs east of house. An interesting old Pear, which I remember with pleasure in farmers' and other old gardens, and now rarely planted. Though small, it has a refreshing flavour.

Sowed down field in Mill Place with the common Furze for use as fodder, and also for covert in a field of not good soil lying against Hastings Wood. Sowed also in nursery lines elsewhere the 'Foxtail' Furze and the Dwarf Furze, two kinds of which seed was obtained in France. Sowed field on 5th April. 1890 *Furze.*

Sowed this wild plant along freshly made fence-banks near new drive and along edge of Bushy Wood and elsewhere. *Fuller's Teazle.*

Many of the fields in Mill Place having awkward or unmeaning names, renamed them on the 7th April, using simple expressive names, and abolishing names descriptive of the mere area as 'the eleven acres', &c. New names adopted: Beeches, Birchfalls, Willows, Hollies, the Grove, Bushy Mead, Brookbanks Mead, Mill Place Mead, Rocks, Upper Rocks, the Hills, Long Mead, Dean Pasture, Oak Flat. *Fields renamed.*

Planted the Warren's field lying above and to north of house with this fine Pine, little known or planted in England. The field, though a light one, lay a little out of the way for the present situation of yard, and I thought it best that the old house should have a wood of some fine hardy Pines above it and on the cold side: the effect from distant points will also be much better. Planted the whole field with stout little plants of the Calabrian Pine, two years transplanted: the tops small, the roots large in proportion. They were obtained from the estate of Les Vaux, Salbris, Loire-et-Cher, France, and were a fortnight or more on the way—10,000 of them in one huge basket. Though nicely packed they heated towards the top, and some were lost through this, which might have been avoided if somewhat smaller packages had been sent. Began planting on the 5th day of April. This field is to test the value here of the way of planting little trees instead of the expensive way of using large and costly plants, 2 to 3 ft. high. The cost, carriage, and labour of planting these is so heavy that no forestry is possible, from an economical point of view, with *Calabrian Pine.*

1890 — *Calabrian Pine.* us so long as it obtains. My 10,000 little plants did not cost me as much as a few hundred larger I had from nurseries about the same time. The point of importance, however, is that the little plants do much better than the big ones—some kinds refusing to transplant well if allowed to get as large as is usual in English nurseries.

Walk beneath Elms. Formed in October so as to give a long simple walk for those who might not care much for walking in the grass. The old walk ran round the wall and close under the Yews. It is much simpler now as it affords an excellent view of the house and a good walk without drip, some of the boughs of the Elm that overhung being removed. The pitch of the ground was altered a little so as to allow of the surface water passing gently off the ground about here. The effect was excellent as soon as done.

Paving of Little Front Garden. To get rid of gravel in the small walks here and also edgings, the necessity of rolling and some other inconveniences of gravel, I began the paving of the little south flower garden with old London flag-stones, and while doing so reconsidered and improved the shape of beds and borders, and got a better effect. The result of paving was very good. Wished to lay the stones at random, but had some trouble in getting the men to do it. These paved walks form their own edging, and it is so easy to change the soil of the bed with stones between. It is a good way in a little garden near a house like this where all the space is given to beds except what is necessary to walk upon.

Azaleas. Planted a variety on the slope above the flower gardens this year so as to plant the bank in a shrubby way. This plan at the same time gives us nice spaces for pretty plants beneath and between the Azaleas. Used mostly the garden kinds, but among them were put nearly all the American Azaleas from which the garden or Ghent Azaleas have sprung.

Heath Garden. In taking away some poor trees on the bank west of house, seized the opportunity to make a bold bed of Heaths in place of an awkward bank. Mostly the finer, not British, Heaths, such as *E. Codonodes*, Mediterranean Heath, and some of our rarer native kinds, were planted. 1890 — *Heath Garden.*

Stone Farm. Purchased for the sake of securing the fine group of rocks in the Upper Rocks Field. Having bought this farm in the Spring, removed all the trees that concealed the rocks and threw them into the same line of view as the rocks in the next field, with good effects from many parts of the estate.

TERRACE AFTER REMOVAL OF GRASS BANK

Holly, Yew, and Ivy by Smugglers' Lane. Seeing that these native Evergreens had done pretty well under the Oaks between the Moat and the Manor House, resolved to add to their numbers. Cleared the lane of other underwood and prepared to plant more Holly and Yew. Cut down a group of Alders between the Moat and the upper pond, these stopping the view, and planted a group of Holly in their place. Made at the same time a rough path through the lane from the pleasure ground or from the Moat. Done in severe weather at Christmas and after.

1891

Late planting of Trees.

Planted the upper part of Warren's Field with Scotch Firs at the end of April; brought them from Woking in Surrey. In Britain the privilege of planting throughout the Winter has been rudely withdrawn this year. For many weeks one could not lift a young sapling owing to the frost. Last year, planting went on all through the Winter, as it so often does in England, barring the occasional week or two of frost we usually have. Planters are sometimes compelled to plant when they can, and the result is often happy. At one time every one planted Evergreens in Autumn or Winter. Now we plant them in May and June, and get a far better result than we could at any other season. The reason of this it is not worth while to discuss, the important fact being what it is. In past days, 'physiological' and many other so-called scientific reasons were given by the 'best authorities' to support the usual and the wrong way. That it was the wrong way was slowly proved by nurserymen compelled to transplant their own stock after the busy season of supplying the public was over. The only difficulty is that in bringing Evergreens from a distance we run the risk of their roots getting dry on the way—a risk less likely to occur in Winter. In planting or thinning home-grown things, this danger need not occur; but care should still be taken, and the roots never allowed to get dry. There are few places, however, where much transplanting of home-grown trees and shrubs might not be done in May. People plant so thickly that it is far better to thin out. If we neglect to thin, the shrubs hurt and in time kill each other. There remains the important question of summer-leafing or hard-wooded trees, as distinguished from evergreen trees like the Holly. The summer-leafing trees and shrubs (Maple, Plane, Ash, Chestnut) are best planted in Autumn or Winter; but it is by no means essential to plant them at these times. I have planted at Gravetye many thousand Hazel, Ash, and Spanish Chestnut in April and May without losing any. The fact that over a large part of Central Europe planting is only done in Spring is not without its lessons for us in hard seasons. In lowland Bavaria and other parts of Germany, where the ground is frozen hard and the snow lies long, planting is always done in late Spring—April to the end of May. As there are far larger and better managed forests in these countries than in England, we have proof of the wisdom of this plan on a large scale. Mere shrubbery planting can usually be done with such care that it alone would be little proof of the value of late planting for woodlands. Where very light dry soils occur, like those in parts of Surrey, late planting will not be so safe; but our soils have rarely this dry and 'burning' nature. They are usually moist enough to make late planting safe. In very cold stiff soils late planting is sometimes the best way of all. In the case of 'specimens' frequently moved, as in well-managed nurseries, or taken up with 'good balls' in one's own grounds at home, the season is not so important a question, and as the labour of moving these is often heavy, it is best to do it when men can best be spared. But even in moving Hollies with good balls, we have a distinct gain in moving them in May—the plants 'take' at once, and growth goes on rapidly, which we do not always find in moving Hollies in mid-Winter. These, on the contrary, often remain almost stationary for several years. Specimen summer-leafing trees are moved with perfect success in Autumn, when in full leaf, and also throughout late Autumn and Winter. I had much loss of young trees and covert plants, like Broom, through being sent long distances without mats or straw round the roots, and being caught on the way by drying March winds. Of feebler plants of Broom and Furze picked up out of the fields where

1891

Late planting of Trees.

1891

Late planting of Trees.

they were 'self-sown', we lost none. They were planted at the same season as better rooted and stronger covert plants that had spent several days on the rail. The success came from the home-raised plants being reset in the ground the same day they were taken up.

Canadian Poplar.

Planted about a thousand trees of the new Canadian Poplar in the Tile Barn Shaw to take the place of many Oaks we had cut there, and also to vary the woods. At one time the timber brought almost nothing, now it realizes a good price. The variety of wants in manufacturing districts makes the Poplar a saleable wood. The round of English wood planting is a little too monotonous. It is not easy to get beyond the Larch, Fir, and a few other things ordinarily planted, without running to the opposite extreme, and planting things useless, or unfit for our climate. A number of coniferous trees that have been introduced and much praised have no proved value; but we need be in no doubt of the hardiness, free growth, and value of the best Poplars. Our native hard-wood trees are so good, that people seldom think of varying them; yet, if it were only for the different colour of the foliage, it would surely be worth planting a thousand Poplars now and then; and when we think of the very rapid growth, both of shoots and timber, and the value of the wood, there are other reasons to induce us to plant these trees. Some of them may grow in soils not good for Oak, Ash, and Beech, and this is an important point in a country like ours, where there is so much marsh land. Belts of swampy land, where one would not think of putting any other tree, will grow Poplars well, and their effect is often excellent in such places. It is wise to grow a variety of trees to suit soils not fit for arable purposes, as by that means we can get the best return from land that will not pay for the labour and cultivation. Their striking landscape value, too, is worth naming; it may be seen in parts of Suffolk—in Constable's country at least. As regards kinds,

Black Italian Poplar.

there are a great many, but for woodland planting the choice need not be large. The commonest kind is the black Italian, and very valuable it is. The Abele is a fine tree, rapid in growth, and large; it is especially good for planting by rivers and in places likely to be under water now and then. The more silvery and *Garden* forms of it need not be considered in woodland planting. I like the common Aspen as a tree; but, though not uncommon in underwoods, we rarely see it as a forest tree in England, perhaps because it is only treated as underwood. In the north of Europe it is a large and useful tree. Some praise much the new Canadian; it grows more rapidly than the black Italian, and this is saying a good deal. In France a Poplar called 'Eugènie' is having a great success among planters in the eastern part of the country. The facility of increase of these trees is in their favour, and leads to their being sold at a very low price in quantities of a thousand or over. They are often planted too large; and frequently stocks of them are bought in sales at nurseries which are merely old refuse, not worth the attention of planters who seek the best result. Besides, the planter has no time to waste over the staking which these lanky trees require. I used healthy, well-rooted little trees, five feet or six feet high. These, firmly set in the ground, require no staking, and soon overtake the taller and older plants.

Evergreen Trees and Shrubs for Britain.

The very hard winter of 1890-1 has given cause for thought to many who care about their gardens! At Gravetye the sweet Bay was cut down in November while in luxuriant growth, and Laurels in the valley were killed at the same time. Although planting in a southern county, I had not trusted to Laurustinus, Arbutus, or any Evergreen frequently killed in Britain, or I should have suffered much more. Even when Evergreens are only killed to the ground, it is a serious disfigurement to pleasure grounds; and it is worth while inquiring if such destruction may not be avoided, and whether we do not

1891

Evergreen Trees and Shrubs for Britain.

pay too dear for our love of Evergreens. So long as they are really hardy, there is little to be said; but most of the exotic evergreens in our gardens are not hardy, and may perish in a severe winter. If it be true that our northern position entitles us to winters at least as severe as the past, were it not for the influence of the Gulf Stream, there is another reason for reflecting on the very common plan of trusting so much to Evergreens in the garden. The trees and shrubs of colder and more northern worlds are as beautiful as any Evergreens—often more so. But when, as this year, in the Paris public gardens, and in many an English garden too, the Evergreens are a mass of dead and brown shoots, even the least observant may see the value of the hardy summer-leafing trees, that take no notice of the frost—Azalea, Weigela, Lilac, Pyrus-thorn, Spiræa, and many others. We are too often tempted to put Evergreens in very important positions. If the Portugal and other Laurels, Bays, or Arbutus, come to grief, we may have ugly banks it will take years to fill up well. Happily, there are Evergreens we may depend upon, and which are quite capable of giving us much of the effect we need; but they do not always get the place they deserve. Soft, half-evergreen Privet, Laurustinus, Arbutus, Euonymus, and tender Rhododendrons are far too much planted. I do not mean on the coast line, where the tenderest may escape, but through the country generally, and especially in plains and valleys. In all cases our own native Evergreens will help us well—Holly, Box, Ivy, Yew, Juniper, not to speak of the hardy Pines. The varieties of our native Holly are an evergreen flora in themselves. I have planted some 5,000 within a few years at Gravetye, but the rabbits and hares, in spite of all precautions, barked many of them this year. Hollies have so many good uses, that in using them in various ways we could almost get all the effect of Evergreens in the landscape that we could wish for. They form the best of hedges for garden or farm.

1891

Evergreen Trees and Shrubs for Britain.

Sometimes we see old mature Hollies forming a fine shelter round a field, and wish young seedling plants were oftener put out with that view, as a line of unclipped Hollies along the cold sides of a field would be almost as good as a shed for stock. There is no part of an estate, from the depths of the woods up to the walls of the house, where the Holly may not be planted with good effect. Perhaps the most neglected of our native Evergreens is the Box; for, though often planted, it is mixed too much with other Evergreens, and rarely forms the beautiful feathery mass it is where it grows wild and alone, as it may be seen near Leatherhead. The Box will not, like Holly, grow on all soils, but it will on most. Covert planters should not neglect it. A big patch of it here and there, where the soil is dry and chalky or sandy, would be excellent. As it seeds freely on warm soils, it would be best to raise and plant seedlings, as the divided plants of nurseries often belong to the edging Box—a more compressed form than the wild variety. Pleasure grounds, too, should have groups of Box, so as to get the beautiful plume-like effect of the branches of Box when growing by itself, and in the full sun. The Ivy generally spreads abundantly, but more should be made of it than is usual in gardens, and less fear of it shown when it climbs trees. The varieties are many and precious. Often in a country place the more common native Evergreens are found together —Yew, Hollies, Ivy. In such a case we may do much to help the effect by cutting away common underwood and by removing a few trees of trifling value, so as to give fuller expression to the Evergreen effects. A little planting of Holly or Yew will often be right in such places to extend the native evergreen growth. The poisonous nature of the Yew when eaten by cattle compels us to guard against it in fields, but there is no reason why it should not be more frequently grown in woods and well planted in pleasure grounds. In these we see the Yew neglected, and big holes of richest loam being prepared for

1891 Californian and other Conifers that never really thrive with us, and some of which are not even hardy. Yet there are Yews in England a thousand years old; and, with the perhaps single exception of the Cedar of Lebanon, it is the most beautiful tree of all that are hardy in Britain. But to deserve that term the trees must be old, and show the handsome and finely-coloured stem. We are not now speaking of the garden varieties of the Yew, but of their old parent, which is worth the whole family put together. In gardens the Yew is only used as a part of the mixed shrubbery, and gardeners rarely see the full beauty of the tree. There are few of the Evergreens introduced from other countries that are as good as those mentioned above. One of the best is the evergreen Barberry of the Rocky Mountains, which will endure a great deal of frost; and no wonder, considering the rigorous winters of its home. It is to shrubs from such countries that we should look for likely things, rather than to those from genial climes. The 'hardy' forms of Laurel, which have been much praised, as compared with the old kind, are not to be depended upon; I saw them hurt this year near London, and at Paris and Turin. At Turin there are handsome trees of that finest of evergreen flowering trees, the evergreen Magnolia. This in England usually adorns our walls only; but it is excellent in all ways. It must be very hardy to winter at the foot of the Alps; and only our want of sun is against it. The popular evergreen Euonymus is not hardy, except by the seashore, and was killed at Gravetye this year, as were Aucubus, Furze, Escallonia, and New Zealand shrubs. There are many tender Rhododendrons, as many will find out this year. People should seek the hardy kinds, among which will be found many raised from the hardy American species. The two hardiest American Rhododendrons (*maximum* and *catawbiense*), the two wild kinds, I planted in large groups as covert plants, and do not see any dead among them as among the garden kinds. A precious Evergreen is the Kalmia,

Exotic Evergreens.

Evergreen Trees and Shrubs for Britain.

a native also of the mountains of the eastern part of America, and therefore prepared for cold. The Evergreen Oak, to my surprise, was nearly killed, and I think I saw it quite killed in the valleys in Kent. I should not again, except on a favoured coast line in England, plant the Evergreen Oak, or any other tree or shrub I have known to be killed. The great means wasted on tender or only half-hardy Evergreens would do infinite good to our gardens if bestowed on trees and shrubs, natives of the northern world—Europe, Asia, and America—summer-leafing for the most part, and quite hardy in our severest winters.

Wood Fires best.

For many years the progress of the British kitchen-range has been triumphant; yet I doubt if a greater fraud exists. People do not always note its cost as compared with other ways, or they would soon be alarmed at its rapacity. But its great initial cost and its cost to keep are not its only drawbacks. It is enormously dirtier than the old fires for cooking, still to be seen in countries where good cooking is understood. During the past Winter we noticed (around a country house, quite isolated and only partly occupied) that the 'blacks' from the fires were visible on the snow for many yards. One may judge, then, how much filth is given off by coal fires even in country houses. In one of the largest houses in England the smoke from the kitchen range in recent summers became a serious nuisance, entering the house again by the windows, and doing its usual work on dress and furniture. The advice of one of the most experienced men in London was sought, and his proposals carried out, but without good result. The cost of fuel is great, especially for those who live in the south of England, a long way from the collieries. To heavy railway charges we have to add carting for several miles, while all the time we may have better fuel for cooking lying useless in woods near at hand. The best cooking can only be done by means of wood. In the best restaurants and hotels in Paris the roasting of poultry and

1891 game is done before wood fires, and for these roasting fires there is no substitute in the multitudinous inventions in ranges and gas. In the south of France fowls are invariably roasted before a wood fire, and in nowhere else have they so good a flavour. An acquaintance in the Bordeaux district tells us that the best form he knew of an English cooking stove was placed in his house there some years ago, in the belief that it was a great improvement on the old French way. It was found to be wasteful and a nuisance, and the best cookery was not possible with it. He has accordingly gone back to the old ways of roasting before a wood fire, braising and stewing and boiling over charcoal ones. Those who notice how things are cooked in Italy and the East need hardly be told that it is a much simpler way than ours, costing much less, and less trying for the cook. An Italian will cook for an hotelful of people with one-tenth the fuss, soot, and cost that is required by one English family. And this cookery is adapted not only for the people of the country, but for English and other visitors, who find the cookery all they could desire, though probably many never think of the simple way in which it is done. It may be that in towns there is no help for our coal cookery and coal bills, but it is not so in the country, where there is scarcely a small estate in which there is not wood of little value to the timber merchant. Cordwood is now very low in price; even if one does sell it with the timber, it may be bought on the ground for very little, or it may be reserved in the sale. Where the timber is cut by the estate workmen there will, of course, be plenty of wood in the best state for burning. The common wood of the south of England, the Oak, is the best for cookery purposes. Underwood has in many districts become so low in price, that it would often be more profitable to convert it into firewood for home use, and to make charcoal of the rest. The ease with which owners of woods may get their charcoal should encourage them to use it. We know people who have to buy

Wood Fires best.

their charcoal, and yet find it better to do so than to endure the waste, heat, and varied filth of the kitchen-range. Where kitchens are not, as they ought to be, on the cool side of the house, the heat of the huge stove is a great nuisance, the heat of the wood and charcoal fires being pleasant in comparison, and endurable in the hottest weather. Therefore it seems to us that, even where cordwood fuel is not plentiful in country places, it would often be better to cut the largest underwood for the necessary cooking fuel, and to convert the rest into charcoal. If in woodless tracts like Egypt, charcoal is brought from Syria for cooking, why lose our opportunities of making use of this good fuel at home? When underwood brought from £15 to £20 an acre, or even more, it was a different matter. Now it is often sold for a tithe of this price, and it would be better to make home use of it. But in other countries not so well supplied with coal as our own, wood is still burnt on the hearth in the right and simple way. The grate is a delusion in every way for the country house or any other, for coal even can be perfectly burnt on the hearth in a well-constructed fireplace. The one drawback is the cost of cutting up wood to fit into the smaller class of fireplaces. Where there is steam or other power this cost is very little; in places where only hand labour is used, something may be gained by the use of the American saw contrivances for rapid cutting of cordwood. I have no experience of these, but hear a good account of them from a German friend who has much forest land. By having a good large boiler we are able to use cordwood whole, a great gain, as even those who have no wood of their own can often buy cordwood at a very low rate in wooded districts. The old way of having a big hearth fire in the hall should never die out, and it should be big enough to take cordwood, i.e. stout branches four feet long. With a good big fire in the middle of the house, and the passages and larger rooms partly heated by hot-water pipes from a boiler, wholly

1891

Wood Fires best.

1891 — *Wood Fires best.* fed with cordwood and a little coke, there would be no need to bring coals from Newcastle to the most distant parts of Southern England, or to wooded districts anywhere. The ignorance of architects about chimneys is so great, that scarcely a house is built without the owner suffering in purse or in comfort, owing to the mistakes made in chimney construction. We do not speak of what Lord Grimthorpe calls 'the young man from the office', but architects of experience. One of these, considered among the best, made at Gravetye chimneys nine by fourteen inches to burn wood in a country house. No good wood fires can be long burnt in such a chimney; it is far too small, and ill supplied with air, as it usually is, it is sure to smoke. As, however, a house in our country is not habitable without good fires and chimneys that never smoke, we advise every one who builds to submit an architect's plans and measurements of chimneys to a chimney expert who has had to do with fires of all kinds. The worst mistakes are made every day, particularly as regards the size of kitchen and other chimneys, leaving much work to do, and making sweeping an expensive and constant labour. No man knows anything about chimneys who, being free to build them as he likes, makes them so that they smoke under any circumstances of wind or weather!

Good Woodwork. There are few things about country houses worse in effect and construction than what is called 'rustic' work. It is complex and ugly as a rule—'its only merit' being that it is so badly put together that dry rot destroys it in a few years. It is perhaps never quite so bad as in garden chairs and summer-houses—things that need not be ugly; but we often see a rustic bridge about as bad as it can be in effect. It is almost impossible to get a garden chair that is not an eyesore, though there does not seem any reason why it should be so. The question, however, is of rustic bridges over streams, on which there is much waste of labour and a very poor result. The prettiest bridge over a stream I have seen was in Wales, in the Cader Idris district, on a farm with a swift stream running through it. To get across some one had cut down a tree that grew near, and chopped the upper side flat, putting one rail alongside. Time has helped it with Fern, Lichen, and Moss, and it was far more beautiful than we had ever seen in any carefully designed rustic bridges. It is not only the far prettier effects we have to note here, but the advantage in strength and endurance. It looked very old and mossy ten years ago, and no doubt it is there now, as the heart-wood of stout trees does not perish like the sap-wood of the 'rustic' work makers. This is no doubt the earliest way of bridging a stream; it will always be the best. It cannot please those who want to design everything, and for whom there is nothing good without a pattern upon it. But it is the strong and artistic way. I have lately made a bridge of this sort at the top of the upper pond by cutting down a stoutish Ash which grew near, and was not in good health, owing to the water encroaching on its roots. The effect was quite good from the first day, and the bridge will last for many years. In all country work true economy is the enduring way, and though we cannot all get the green stone gate-posts, stained with yellow Lichen, of the farms about Keswick, or the lasting granite fence posts one sees in Italy, we should strive against makeshifts and things being done over and over again. Foot-bridges they should be called, as they are too narrow for cart traffic; but with a good Oak rail at one side they serve every purpose in pleasure grounds or woods, or for handy ways for men crossing streams to attend to stock. This tree bridge is distinctly better than planks for bridges over streams where there are rights of way; planks rot soon, and there is generally little time to look after such things; accidents sometimes happen, and as the owner is responsible for the repair of foot-bridges along rights of way, it may be awkward for him. Rustic bridges strong enough for much farm or woodland carting may easily be made from

1891 — *Good Woodwork.*

Foot-bridges.

1891 home-grown materials. Two Oaks across for the main supports, a layer of Larch stems across them, laid close, tops and bottoms alternating, so as to form a close bed. On this strong foundation grass turf is put face downwards, and then a bed of any stone at hand, to make the bridge firm and level with the roads leading to it. Protected by strong Oak rails it looks well, better than more pretentious constructions. It is far better to work in this way with tree bridges, except where stones lie about the ground, and where there is cheap labour to work them up. Stone put together in a strong and simple way is best of all. Where stream beds are rocky and shallow, stepping-stones are often prettier and better than any bridge, but they are useless in the many streams cutting through alluvial soils and with high banks. Where stepping-stones are practicable, they should be used. As to seats there are few who have skill to make them well and simply, and there is always the decay of wood to be considered. If any of our own woods be used, Oak is the best. A very simple iron frame without any ornament is best, with stout heart-wood Oak in strong laths screwed to it. The Oak is best not painted and should be allowed to take its silvery colour. There is no bench seat so good as a marble or hard stone one, of simple design, and always strongly made. In our country one objection to stone seats may be met by a mat or a light trellis of Bamboo, or other wood, placed upon the stone. Now and then I have formed good stone seats from steps and other stones displaced in making building alterations. Stone seats should be always set on stout stones, bedded in concrete. The rustic trellis covered-way may be a charming thing in a garden, as it forms such a home for climbers, besides affording us a shady way. Shade is more essential in other countries than in ours, and the Italian covered-way is often very picturesque. The best material for the supports of a covered way is the Italian one of rough stone. With lasting supports like these, the woodwork is easy. These simple rough posts

Good Woodwork.

The Covered-way.

may be had in certain quarries in England, as in the Lake 1891
country; but where they cannot be, it will often be better to *The*
build a brick or rough stone pillar than to trust to any wood *Covered-*
enduring in the ground. Where wood is used, Oak free of *way.*
sap-wood is best; Larch, good; and best of all, perhaps, the common Locust tree, which, however, is not often plentiful

IN MILL PLACE FARM

with us in a mature state. For all other parts of covered-ways nothing is better than old Oak branches, or stems of stunted Oak, or old stub Oak, that often occur about a country place, and are of very little value to the timber merchant. Larch lasts well and may be used in the absence of Oak, but it is not nearly so good in effect. By the use of Oak with stone or brick pillars a covered-way for climbers may be made to last for

1891 — *The Covered-way.* many years, without falling into the decay so common in this kind of work when done with wood that soon perishes and without lasting supports. Better far use strong iron wire than wood of this sort. The great advantage of the well-made woodwork over the iron is in effect, as, carefully done, a covered-way made as above described may be picturesque even before there is a plant on it. After covered-ways, seats, and bridges, perhaps the most essential thing in woodwork is the fence.

Fencing. The presence of the iron fence destroys the beauty of half the country seats in England, and this evil is growing every day. There are various objections to iron fencing, but we will only deal with its effect here. This is as bad as it can be. Any picture is out of the question with an iron fence in the foreground. Where an open fence is wanted in a place with any landscape beauty, there is nothing so good as a split Oak post-and-rail, made, as it ought to be, of good heart Oak, and with stout posts. Fine in form and colour, an artist would be glad to have it in the foreground of his picture, whereas the iron fence drives him away to farm or roadside. Any sawn wooden fence is neither good nor pretty. As on many estates Oak is plentiful, good examples of split Oak post-and-rail fences should be far more common. The Oak park fence is common, and, while often the best thing that can be used, we sometimes see it where a good live fence of Holly, Quick, and Wild-Rose on top of a good bank would be much better.

The Summer-house. Lastly, we come to the summer-house, generally a failure, and often a heap of decay. Making such structures of wood, that soon decays, is wasted labour. Where a summer-house is really needed, it would generally be better to build it of rough stone or some enduring material, and cover it with vines or quick growing climbers. As such houses are generally small, it would be often the best way to put a stone roof on them, sloping gently to the back. It is possible to make a pretty summer-house out of live trees, if not out of dead ones. An old Yew, or a group of old

Yews, or low spreading Oaks (there is a fine example of this in the pleasure grounds at Shrubland Park), an old Beech, or a group of Evergreen Oaks, make the most delightful self-roofing summer-houses. With a little care as to aspect, and trimming of old and usually worn-out branches, so as to get air and room below the trees without injuring their beauty, it is easy to form pleasant bowers for hot days. 1891 *The Summer-house.*

Planted on July 3 some of my favourite Apennine Anemone at the bases of the Oak-trees in the grove of these between the house and the Moat. They seem to be happier in the mossy and leafy soil that gathers at the base of old trees. The warmth and shelter of the sunny corners and the tree bases also helps them in our climate, but this Wind-flower is more free in limestone and free soils. *The Apennine Wind-flower.*

These fields thrown together and simplified in outline in the Autumn of 1892; lane also removed in same year to extend and improve the Furnace Field, November, 1892. *Crowsfield Farm purchased from Vicar of Portslade, 1891.*

Took down fence between the two fields at Mill Place known as the Middle Bottoms and Heather Bottoms, now called Brookbanks. Took down fence between 'Little Field' and Mill Place Mead. Grubbed shaw and fence between the Furnace Field and part of Crowsfield, now called the Furnace Field. Took away shaw running up to Crowsfield Cottage from Mill Place, and grubbed and levelled the ground here—a heavy work. *Fields simplified.*

BOY

1892

Clearing Upper Lake. Cleared the upper part of this, for the second time within five years, of mud. Surprised at an accumulation of 18 to 24 inches of silt, decayed leaves, &c., brought down by the ever-busy little stream. This pond being of recent formation the edge was always hard and bare from the absence of the vegetation natural to such places, the water beating an ugly line between itself and the grass, so in clearing it we left a fringe and the rich stuff along the edge—even adding to it here and there, so as to encourage a line of handsome plants between the grass and the water. Purpose to set Water Lilies, Flowering Rush, Bulrushes, Sweet Flag, and all the nobler Reeds and Water Flowers to form a beautiful fringe round the edge. This will greatly improve the effect of water-side.

Bullace. Being interested in this native fruit I got what I could from different sources, as I supposed, both the yellow and the black Bullace, with a view to testing its value as a fruit, and also in the belief that it would be pretty. Some put in are near hedgerows, others are planted as standards; some were got in the district, others in Kent, the black Bullace from Colchester and some from Dorset *.

Trout. In the Lower Lake the carp being very numerous and nobody being able to catch them, I had to have them netted out as they were considered injurious to the trout by eating the spawn and fouling the water. We got out as many as we could. Had eventually to lower the lake to get rid wholly of the carp.

Goose-berries. Planted a great number of the best eating kinds—the small and medium sized berries of fine flavour, which are so good

---

* Note, 1910. *Never did any good: soil against them.*

in Scotland, whence I get my stock. There is no native or cultivated fruit quite so delicious as the best kinds of Gooseberries, certainly none which suit our climate so well. 1892 *Gooseberries.*

Note on Growth of Oak in Shagswell Wood.

The following table, drawn up by the late Mr. W. Stenning of East Grinstead, shows the growth in the value of Oak, &c., trees and underwood in Shagswell Wood, West Hoathly, in the period mentioned. That the common idea that Oak is a slow growing tree is wrong is shown by the following table: *Growth of Oak.*

1858, *May.—Standing in 16-Acre Wood.*

	£	*s.*	*d.*
6 oak, 20 ft. and upwards, 130 ft., at £6 10*s.* per load	16	18	0
48 oak, 10 ft. to 19 ft., 547 ft., at £5 10*s.* "	60	3	4
171 oak, under 10 ft., 1,084 ft., at £4 10*s.* "	95	3	2
6 ash, 18 ft. "	1	4	0
351 tellars (1*s.* and upwards)	30	12	0
	£204	0	6

1868, *April.—Marked for Sale, and subsequently cut down.*

	£	*s.*	*d.*
1 oak, 43 ft., at £6 10*s.* per load 11 oak, 20 ft. to 39 ft., 286 ft., at £6 10*s.* per load	42	15	5
91 oak, 10 ft. to 19 ft., 1,236 ft., at £5 15*s.* "	142	0	9
179 oak under 10 ft., 910 ft., at £5 "	91	0	0
4 ash, 36 ft., at 1*s.* 6*d.*	2	9	6
85 tellars	6	10	0
	£284	15	8

1868, *May.—Standing.*

	£	*s.*	*d.*
180 oak, 804 ft.	81	13	4
529 tellars (1*s.* and upwards)	36	3	0
16 ash, 46 ft.	3	9	0
	121	5	4
Sold as above	284	15	8
	£406	1	0

The underwood produced £9 per acre.

1892

Purchase of Old Coombe Farm. Purchased this property from Messrs. Stenning with a view of adding the considerable woodland to the large mass already acquired, and also because it brought the land up to West Hoathly Station, giving us a road to that station which would enable timber and underwood to be more easily cleared. It was free of land-tax and great tithe.

Water-lilies. The first brought here and planted in the Upper Lake were the common white and yellow from Shrubland Park; white from Burghleigh, and the American yellow Nuphar from the Manchester Botanic Gardens. The ponds, in their present state at least, being of recent formation, there were few water plants along them, and no Water-lilies or even Forget-me-not on my coming. Planted also a new hybrid Water-lily, raised in France (*N. Marliacea*), in the same pond.

The Rocks, Stone Farm. Having long thought that the best thing for these would be to leave them exposed in the field, it occurred to me this year that it would be better to fence them and the ground near with a view to planting a beautiful series of rock shrubs and wild Roses. I began to make an ordinary bank and dick fence with wild Roses set between the sods, in early November, without attempting a very choice rock or Alpine collection, which might tempt people so far from the house. I hope to be able to plant it that it will be at once very beautiful, and a pretty retreat for winged game. The smaller and prettier rock and mountain shrubs I should make a special feature of, with the Alpine flowers most easily grown and bulbs loving warm sandy soil, which we have not near the garden proper.

Exhibition of Pictures made at Gravetye in 1891. Mr. Moon, after coming to see us for the sake of the flowers and landscape, came with a friend in 1891, and I was much interested in the sketches they made in early Summer. I suggested that they should stay for the season and study the beauty of the place from the landscape painter's point of view, which they accordingly did. They worked at the pictures all the Winter, and eventually they secured a place in which to exhibit them at Mr. Gooden's in Pall Mall. Annexed is a catalogue of the exhibition, which opened in the first week of November, 1892, and was alluded to as follows in the *Times* of November 14, 1892:—

1892 — Exhibition of Pictures made at Gravetye in 1891.

At Mr. Stephen Gooden's gallery, 57, Pall Mall, there is on view for a short time a very charming little collection of cabinet pictures by two artists, Mr. W. E. Norton and Mr. H. G. Moon, who were settled for the summer of 1891 at Gravetye Manor, Sussex, and have painted the house, the garden, and the neighbouring woods and fields in every aspect and under every condition of sun and shade. Mr. Norton is an American trained in Paris, Mr. Moon is English, but there is enough likeness between their methods and points of view to make the collection pleasantly harmonious. We will not attempt to decide the prize of merit between the two artists; *et vitula tu dignus et hic;* both are clear-eyed observers of nature, and each paints with the brush of a thoroughly trained artist.

Catalogue of Pictures

A STORY OF THE YEAR ROUND A COUNTRY HOUSE

That things made for use are often more beautiful than those designed as ornamental is an instructive fact. A farm-house and its buildings are often beautiful, though we know the people who made it only made what they wanted for use. The open country in many districts in England is beautiful, while grounds designed to be 'ornamental' drive the artist away by their ugliness. That beauty and use are one is clear in many ways, but this has rarely been true of the garden and home landscape. Artistic beauty could not arise before the supposed necessities of the art of 'laying out' of grounds, and we have had to tolerate every form of ugliness and formality under the supposition of their being necessary parts of garden design.

The artist has been driven from what should be the heart of the landscape. The carefully set out ground round a country house is often as ugly as it can be, while the open country near may be full of pictures! In our own day, and in all the days

1892

A Story of the Year round a Country House.

we know of, the home landscape has very often been cut off from the sympathies of artistic people. We rarely see pictures of country seats, because the artist finds nothing to do in them except shut his eyes and get out into the open country as soon as he can. The bad example of public places was too often followed. Who could paint the upper terrace of Versailles, or look at the result if any one could be found to paint such a graceless waste? It is only one of many large and ugly places that have hitherto passed as all that was admirable in garden design. With the sky above us, and the earth and living things upon it, we need not make our home landscapes ugly in any way. At Gravetye the house, in a pretty wooded country, had escaped 'improvement' in the days of stucco and the 'Grecian villa'. Owing to the place being long used as a farm the fields and woods were for the most part left wholly for convenience of working and clearing. There was no pleasure ground or park; the woods, made for use, crept close to the house, connected by little 'shaws' creeping through the hollows; these shaws occupying ground which could not be cultivated, it was happily left to Oak and Ash and Hazel, tall Grasses, stout Wood-flowers, Primroses and Ivy. Although almost every field was dealt with, beauty was never lost sight of. Nothing, from making a road to forming a fence, was done without considering its effects on the landscape from every point of view. The fields, small in this old forest district, in being enlarged were given good boundaries, often woodland boundaries, owing to the number of woods and shaws on the estate. Any straight hard common fences, cutting a country up like a chess-board, were removed in favour of more picturesque dividing lines, often following the natural dividing lines of the bosom-like hills. The trees in the woods were left to be dealt with from a woodman's point of view, but those in the open fields were carefully considered as regards effect, and confusing or obscuring trees removed. Any planting done was in masses, which worked in with the other masses of wood, and did not produce a spotty effect. 1892

A Story of the Year round a Country House.

In the garden we got plenty of lovely flowers, not set in patterns (oilcloth, tile, or other), but in simple graceful masses and groups; myriads of Daffodils and other early flowers in the meadows near the house in picturesque colonies and groups. In the flower garden we have self-coloured Pansies, as pure in colour as their mountain parents on the Alpine turf in June, many Carnations in their native strong and handsome self-colours; Tea Roses everywhere, not stuck in a formal 'Rosery', and far into the Autumn a sea of Lavender and pale blue Starworts, which, seen here and there in drifts through the trees, were tender in colour as the far azure seen through lacy veils of cloud.

And so, after some years of labour, my artist friends came on the scene. It was in Apple-blossom time, and charmed by their sketches of the Apple-trees and the Narcissus mead, I asked them to stay and watch the beauty of the year on one spot of English ground. This they did, from Daffodil time to the fall, and even into Winter, when the colour in Sussex woodlands is so beautiful. But the Summer of 1891 was wet and stormy, and often the artists, though hardy men, were driven off the fields by heavy rainstorms, which also drove away the pictures. One thing may perhaps be said with profit in a day when there is so much talk of 'idealizing' things. Those who hold such views of the landscape art will find no encouragement here. The aim was, as it always should be, to get as near the visible beauty of things as it is possible for the artist to go. Clear and loving eyes see enough of the divine beauty of sky, sea, field, and tree, ever to wish for more than getting as near a faithful record of that beauty as the winged minutes and ceaseless changes in sky and land allow to human effort.

Hardy Heaths.

Planted nearly all obtainable kinds of British Heaths and other hardy Heaths in large beds in ground north of the House,

1892

Hardy Heaths. intending to let them, once established, form wild masses, i.e. not cultivated much after first planting, and planting hardy bulbs, and things for the wild garden, between them.

Cedar of Lebanon. Bought 1,000 trees and planted them in ground above the stable to form part of wood of hardy pines (aged 2 years). *Riga Pine.* 1,000 trees, and planted in same wood near the pit hole where we dig stone for the stable walls (aged 2 years). *Austrian Pine.* 1,000 trees planted in centre of same wood (aged 3 years). *Corsican Pine.* 1,000 to fill gaps made by rabbits among those already in the same field. *Larch.* A few thousand were planted among the Corsican Pines near Birch Farm.

Winter Jasmine at the Moat House (Jasminum nudiflorum). This flowered charmingly at the Moat this Autumn in very dull weather, and was so pretty on all sides of the little house that I thought it might be well to use it more. This beautiful plant is so well known that it occurs in many cottage gardens, facing our Winter with its delicate buds and flowers. We have not enough of it, and it is not placed in sufficient variety of situation so that we might secure successive bloom. If we put it around the house it ought to be in different aspects and in some quantity, so that if it catches the sun after a sharp frost on the one side of the house it would escape on the other. Its value in the house is considerable as the buds open in water, and it may be gathered in the bud state, and the flowers open unbruised. If pruned 'hard', its flowers are not nearly so numerous or gracefully borne. I propose, as soon as I can get the plants, to plant the Winter Jasmine on all sides of the Moat, to set the old house in it in fact, and also to garland the railings, trellises, and fences near the Moat with the same plant, so that one may see it in many attitudes, and see what effects the plant may give in some quantity near the old house.

Planting Woods. The Corsican Pine in young plantation near Birch Farm having been injured by rabbits, I had to fence the remains of it

1892

Planting Woods.

and replant. The Larch near fared a little better but was also injured. From my experience here no young Pines are safe from ground vermin in a woody district. The wood above the house with a mixed plantation of Firs and some Scotch suffered much, and we had to refill the spaces this year with Calabrian Laricio, Crimean, Austrian, Cedar, and a few Douglas Firs. We also had to wire this, and determined to do so in an effective way, so we gathered experience, and our future plantings will be done differently. I do not mean nurserymen's planting in a garden, which may be anything; shrub planting is generally as bad as it can be from garden or other point of view, and Conifers stuck about like children's toys we may pass by. This note refers to forest trees, mostly coniferous, which attract by their beauty or other qualities. In many districts Oak and Ash and what are called hardwood trees are plentiful, and the question often resolves itself to the planting of the hardier Pines and Larch. There are various ways of doing this, but as it is a costly matter and the result long to wait for, planting should be done with economy and knowledge, and mistakes avoided. A few simple rules for planting trees may be laid down: 1st. To plant a Larch wood we should gain time by fallowing the ground and planting it in Autumn, ploughing the field, and moving it about in the Summer, fencing it with close wire rather early in Autumn, and when ready write in November for the trees. We should thus gain a year and have no blanks to speak of. 2nd. Never plant large plants. Many nurserymen send out the trees too big and with more top than root. Whether we plant from our home nursery or buy, it is better to begin with trees not over a foot high, often less. This Autumn I planted 1,000 Cedars of Lebanon with shoots 6 in. high, and no doubt they will take well and be finer trees than if planted at 6 ft. instead of 6 in. When little trees begin to grow the growth is clean and the roots take hold in a healthy way. 3rd. Never order or disturb trees till the field

Rules for planting.

1892 — *Planting Woods.*

is ready and fenced. As many people buy in nurseries they have to face the accidents of drying winds, long railway journeys, and if not ready when the trees come, and the trees be put in stock until the ground is ready, the chances of failure are increased. I have seen this year many thousand Larch lost from these causes. In 'rabbity' districts it is necessary to wire well. This is best done in late Summer or early Autumn, so that we need only write for the plants if they are not in our own nursery, and it is then a pleasure to get all hands and put them in. In a field so planted there will be few failures. Another excellent rule is never to plant a tree on a large scale not known to be hardy and a healthy grower in the district in which you plant. There are so many novelties offered amongst Conifers that we must be careful *not* to plant woods unless we know the tree will thrive and endure the hardest winters. I was surprised to hear this year, when at Orleans, that the Corsican Fir had been lost in Central France, where it was thought one of the hardiest. We have a variety of precious trees, and may well take the trouble to find out the hardy and enduring. Besides the often planted Austrian Pine and the Spruce, Scotch Fir and Silver Fir, there are some that deserve to be more planted, e.g. the Weymouth Pine, Riga Pine, Calabrian Pine, the Crimean Fir, and the Cedar of Lebanon—this last, mostly put out in gardens as a beautiful specimen by itself, should oftener be planted as a wood. Planting it apart we get a fine 'specimen' and it develops huge side branches, whereas in a wood it grows erect, Pine-like, and is quite as beautiful, raising its plumy head. It is also lighter and more pointed, and there is less danger from the snow, which so often destroys it when spreading on the lawn. One of the most beautiful things is a wood of the Cedar of Lebanon grown, that is to say, as it generally congregates on the mountains. We have proved its hardiness for some generations in England, which cannot be said of many Pines.

Five years ago planted a moist slope in new drive in Warren's Wood with the Cardinal Willow, and ever since it has been a pleasure to see the young trees. In various lights of late the effect of these Red Willows is the most beautiful thing in the woods. In Summer, when the delicate silvery leaves partly hide the red of the stems, the effect against the sky is often beautiful; and now, too, even one Osier tree by the water-side is often a picture. Where planted with judgement in some number the effect may be charming. The Red Willow or a variety of it is sometimes called the Cardinal Willow. The White Willow is also a noble tree, though not so effective in Winter as the Red or Yellow Willow, but in Summer beautiful and the best of all for the marsh. When the gardener plants a Willow it is generally some curious thing with a mop head, like the American Weeping Willow. 1892 — *Cardinal Willows and Osiers.*

Country gentlemen might therefore take the tree Willows under their care, and plant them in colonies here and there, by water, or in wet, marshy, and often useless places. A marshy place planted with underwood formed of the Yellow or Red Willow would be charmingly picturesque, and as useful, if not more so, than some other underwoods. There is no difficulty in getting any of these Willows by the thousand. The Red Willow is oftener seen in French nurseries and I have seen something like it wild in Ireland. In any case the common Yellow Osier should be planted freely. In places much haunted by the rabbit, young Willows of these kinds disappear as soon as planted by streams; in meadows where there are cattle they also perish. So that in certain districts a little care may be wanted to protect them from animals. In the whole range of planting I know nothing which gives more satisfactory results than these forms of our native Willow. There are American and other native Willows that might be planted with them, but none quite so good. An interesting point about some of the Willows is their use for tying, which is often forgotten now.

1892

Cardinal Willows and Osiers.

There used to be many things done in gardens in which these Willow shoots were used for tying and packing, and now we notice wire, tarred twine, and other manufactured articles used instead—a mistake. There are nurseries where millions of young trees are sent away every month in Winter, all tied with Willow. It is not one kind only that may be planted for the purpose; but there are half a dozen Willows which are useful for tying in addition to the common Osier.

WOODLAND STREAMLET

1893

Oak Walk and Oak Boathouse.

Resolved this year to connect the lower lake with the pleasure garden (before cut off from each other by the sunk fence), and to this end made a way from the upper lake to the lower, so that one could get from the pleasure ground from the house to the lower lake easily. Built a boathouse between the little island and the bank entirely of native Oak (including the roof of shingle). I expect it will last many years, and, being all of our own silvery Oak, untouched in any way by paint or varnish, the effect is good. Fenced and wired a portion of the ground along the lower lake and planted a collection of Daffodils (thinking more of the effect of the finer kinds in masses and groups than of rarities. Within the wire and iron fence, which cuts this Oak walk off from the field above it, made a bank and planted it with a picturesque fence of Wild Roses and flowering trees, and shrubs of the Plum, Sloe, Thorn, and allied trees. The idea is to make this part of the grounds mainly a covert of hardy native things. The only planting of much consequence, after the fence of flowering shrubs and trees, was of a collection of Willows and like feathering trees, which we planted on the island and by the banks. On the sides of the path to the boathouse commenced to plant masses of the nobler forms of the common Daffodil, e.g. the finer forms allied to our native Narcissus, such as *princeps, maximus* (or indeed any of that trumpet flowering group), thinking that these would look very beautiful in the Spring under the grove of Oaks running by the lake. This addition to the pleasure garden is made without sacrificing any good land, and gives us an interesting walk and most beautiful views, while forming a fence between the cattle grazing on either side of the lake, which frequently crossed

1893

Oak Walk and Oak Boathouse. over to each other before the fences and coverts were formed. To get from the pleasure ground to the lower lake, we took down the little conduit, through which the water ran from the upper to the lower lake, and made it high enough to form a pathway by adding a stone platform above the water. In this way we get easy access from the south lawn to the lower lake without exposing the garden in any way.

Beds in Flower Garden. Finding the soil used in the original preparation of these beds not as deep, nor the beds as well made, as I had expected (they were made by the ground workmen), we ripped up nearly all of them, and digging out clay and rock from the bottom, made each bed two feet deep below the level of the grass, filling in with good loam, sand, and manure, as the best preparation for Tea Roses and Carnations. Dug up at the same time the great border round the garden and excavated it in the same way, lowering it a little and making it much better for the Roses.

Rabbits. The memorable dry season of the present year, lasting as it did for eight months of pleasant sunny weather, had many interesting effects, not the least noticeable being the immense increase of rabbits, owing, probably, to the absence of accidents to the young from bad weather. These were a harmful pest in the crops and pastures, notwithstanding all our efforts to destroy them. My first plan of wiring the place all round was given up since I purchased the woodlands around, and the plan should not and cannot be carried out except in the case of young plantations, gardens, and the like. In the Autumn, however, the very little feed we had left became so injured by their attacks, that I had to employ a man solely to attend to this work of keeping down the rabbits, which he commenced very successfully in November, mainly depending on snares. In this way we hoped to keep the fields more free of them, as they not only destroy much valuable food, but make what they leave far less grateful to the stock.

1893

Sheep: South Downs a failure.

Sold the flock of South Downs in the Autumn. We had succeeded in getting together a pure strain and with care had kept them in good health; but, as I gradually learned that the ground was not such as would keep them so for many years, and for other reasons, I thought it best to give them up and devote the land entirely to the Sussex cattle. One other reason was that sheep are only profitable when a flock of sufficient size is kept, the expenses of a small one of two hundred ewes being almost as great in certain ways as one of five hundred, at less than which number they are rarely profitable. The great scarcity of food owing to the drought left me no alternative but give up the sheep, and we sold them well considering the prices of stock this exceptional year, though at far lower prices than they were bought.

Thorns.

Planted a great group of Thorns (American and European species) on the rough bank above the Moat House in February, and this position being well adapted to show their beauty, it is hoped they will thrive for many years there.*

Ivies.

Planted large groups of the different kinds of green Ivy along the path from the house to the Moat (January) to grow on the ground in natural masses and run over the banks and up the trees.

Lilacs.

Planted large bed on south lawn with Lilacs on their own roots. Having suffered so much from the grafted kinds, I made special efforts to get the plants on their own roots of the kinds obtainable in that way, either in French or English nurseries.

Peonias on S. Lawn.

Planted a great group of the single species beneath the Crab group on the south lawn, massing together as many plants of each species as we could get.

Clematis (wild kinds).

Planted a number of the single wild species in the hedge-rows and the Rose fence near the garden, viz. *C. montana*, *viticella*, *flammula*, and *graveolens*.

---

* NOTE, 1910. *Many failed through being grafted.*

1893

Honeysuckles. Planted groups of the finest kinds on the margin of the new orchard and in other places, trying to show as much as possible of the beauty of this precious group of hardy plants, so graceful and so hardy and vigorous.

Belladonna Lilies. Planted 100 of these against the warm walls in the little front garden (failed from coldness of soil).

Hardy Bulbs in Grass. In lawn south of house and along the Oak walk at lower lake planted over 70,000 of hardy bulbs such as we knew would be likely to live and do well in the natural turf of the lawn and lake side.

Oaks from Acorns. Thinking it would be interesting to raise a wood of Oaks from Acorns and to study their growth raised in that way, which I believe to be the best, as it is the most natural way, we sowed the 'Grove Croft' in Mill Place with Acorns, which were very plentiful and large, and of which we chose some of the largest—the Oaks in the place showing great variety in the size of Acorns and in their form too.

New Water-lilies. Planted a large collection of new finely coloured kinds raised in France and sent me by M. Latour-Marliac and M. Maurice de Vilmorin. We owe a deep debt of gratitude to M. Latour-Marliac, who has given us an addition to our hardy garden flowers which cannot be over-estimated. He has added the large and noble forms and the soft lovely colour of the Eastern Water-lilies to the garden waters of every northern country. If this merely meant the beauty of the individual flower our debt would be great, but it is not that only; the splendid beauty of these plants will lead people to think of the true and artistic ways of adorning garden waters. Our own Water-lily was always neglected in gardens and rarely effective except in a wild state; but when people see that they may have in England the soft and beautiful yellows and the rose and red of the tropical Water-lilies throughout Summer and Autumn, they will begin to take a greater interest in their garden waters. Even the wretched formless duck-ponds which disfigure

New Water-lilies.

so many country seats will at length begin to have a reason to be, and, apart from Water-lilies, the change will be the means of leading people to think more of the many noble flowers and fine-leaved plants of the water-side. A great many handsome plants frequent the banks of rivers and lakes and are never so fine elsewhere. The rich soil brought down by all rivers is a great aid in the growth of water-side plants, and those who have never seen them in this their natural position have little idea of their beauty or size. Apart from the true water-plants there is also a whole series of noble hardy flowers which should be massed in natural ways by the side of water, like Iris, Meadowsweet, Loosestrife, Globeflowers, Knotworts, and many others, some of them like the Rosy Meadowsweets flowering with these Lilies. In such ways we should get gardens of lovely colour and form free from all trace of the pastrycook gardener. We had great success in the past Summer with these Water-lilies, which flowered till the middle of Autumn, noble buds and flowers being gathered towards the end of October. Some may think that the sunny days of the present year may have been sufficient to account for this success, but among the many beautiful plants Lord de Saumarez has tried at Shrubland are all the hybrids obtainable of this never-to-be-forgotten raiser in the by no means favourable years preceding the present memorable Summer of 1893, and they did well and flowered admirably. Shrubland is a long way north of London, so we have some guide as to the area over which these Lilies may be grown in the open air. Our plants are entirely in the open air, and I do not fear any loss to any of them from frost. The first I had and which I timidly put out, was not in the least affected by the frosts of several years. Most of the new kinds were planted in the Spring of the present year, and in a few months we got very good blooms and red Water-lilies visible at long distances the first year of their growth. In view of the great interest of the subject and of the fact that so

1893 — *New Water-lilies.*

little was known of the way in which M. Latour-Marliac had originated his Water-lilies, I felt that the best way was to apply to him, and he kindly sent me the following interesting account of his precious work:—

'About the year 1879 I commenced the work in earnest by crossing the finest types of the hardy and tropical Nymphaeas which I had in cultivation here. These early attempts were at first negative in their results, but soon afterwards I had an unexpected success in obtaining a hybrid with deep red flowers, the seed-parent of which was *Nymphaea pygmaea alba* fertilized with pollen from the flowers of *N. rubra indica*. To my great disappointment this magnificent specimen proved hopelessly barren and from it I obtained neither seeds nor offsets, so that, after having tried in vain to reproduce it, I gave up the task and turned my attention in another direction. In order to obtain plants of a fine character, I thought that it was essential that I should make it a point not to employ as seed-parents any subjects except such as were very free-flowering, and by vigorously adhering to this principle I succeeded little by little, by means of numerous sowings and strict selections, in raising types which were in every way improved in the form and characteristics of their flowers. It was thus that one of these new subjects, *N. alba*, fertilised with pollen from the American species, *N. flava*, produced *N. Marliacea chromatella*, which has achieved such a high reputation. In the following year I obtained the hybrid *N. odorata sulphurea* from a similar crossing of *N. odorata alba* with *N. flava*, and the last-named species has also been the pollen-parent of *N. pygmaea Helvola*. About the same time two species bearing a high character made their first appearance in gardens, viz. *N. sphaerocarpa* (a native of Sweden) and *N. odorata rubra*. I determined to reject the first for hybridizing purposes and I gave all my attention to the fascinating American species *N. odorata rubra*, which I employed as the pollen-parent with my choicest specimen of *N. alba* as the seed-parent; this alliance rewarded me with the sweet *N. Marliacea rosea* and *N. M. carnea*. *N. odorata rubra* was subsequently the parent of the beautiful *N. o. exquisita*, the colour of which is pink approaching to carmine. As the last raised specimen of this first group of my hardy hybrid Nymphaeas, I must mention the remarkable *N. M. albida*, the flowers of which have not yet been surpassed in size by any other *Nymphaea*. In the year 1889 the Universal Exhibition was inaugurated at Paris, and my small collection of the above-named hybrids timidly took the road to the metropolis. Their graceful elegance was appreciated and they came back with the distinction of a first prize. What a change since then! And with what assurance would that first collection have made the journey to Paris in company with the splendid race which has since made its appearance! Their success at the Universal Exhibition encouraged me to further advances, and I applied myself to the object of effecting a cross which should produce plants with bright-red flowers much finer in colour, to *N. sphaerocarpa*. *N. o. rubra* proved incapable of supplying the desired improvement, and after numerous trials and experiments I attained my object in a hybrid of the same colour as the tropical

N. rubra, that having also the property of bearing seed all the more precious because it produces no offsets. As I anticipated, the plant could not be sent out as the seedlings could not be relied upon to resemble it; and in fact it has produced seedlings with flowers of intermediate shades, from soft pink to deepest red. Those varieties, which it is impossible to render permanent, owing to the failure of their stems to yield offsets, have proved very useful for hybridizing choice varieties of the stoloniferous and proliferous kinds; and from these I have obtained some remarkably hardy novelties which for six months of the year embellish the waters of pleasure grounds with their splendid flowers. Most of them are catalogued as *N. Robinsoni*, *Seignoureti*, *Laydekeri*, *rosea*, *liliacea*, *fulgens*, *Marliaceaignea*, *N. M. rubra punctata*, and *N. M. flammea.* Others no less brilliant will soon be added to the list. The acquisition of a red-flowered hybrid yielding seed has opened up a new prospect by affording means of effecting crossings with the yellow kinds, the result being quite a legion of Nymphaeas with flowers exhibiting singular shades such as orange, vermilion, gold. Some of these have been introduced to the public, the first to flower being named after the editor of the *Garden* in return for the great interest he has taken in the advancement of the culture of the hardy Nymphaeas. The blending of Nymphaeas of the Castalia tribe of northern countries with the Lotus of the tropics is now a fact; but there remains the important task of hybridizing the Castalias with plants of the Cyanea section, which includes a number of superb blue-flowered Nymphaeas, a work well calculated to stimulate the enthusiasm of the hybridizers.'

1893 — *New Water-lilies.*

Pine Wood North of House.

In December added to young wood of Pines north of house (formerly Warren's Field), and some more of the hardier Pines to replace some that were destroyed when small by rabbits, and also to vary the wood a little. Among these additions were the Swiss, Crimean, Weymouth Pines, and the Colorado Douglas Fir. Looking for good Pines in catalogues one is struck by the numbers of useless Pines which are offered, not only the tender trees of California, which in their own country are beautiful, though unhappily not so in ours, but also the mass of distorted, unnatural, and ugly forms, the names of which disfigure the best catalogues. Of all the Pines which have been brought to England and become popular, the worst failure is the Norway Spruce, which is everywhere in Sussex and Surrey in a diseased state, save by streams or in wet bottoms. The Norway Spruce where it does well is a beautiful tree, but in most cases in the Home Counties it will not grow well, although good specimens may be seen here and there in moist

1893 soil. But what of its wretched varieties—dwarfed, compacted, starved, inverted? In one foreign catalogue there are no less than twenty-eight varieties of the Norway Spruce in all sorts of dwarf and monstrous shapes, two of them being even disfigured with the name *monstrosa*! The true beauty of the Pine comes from its size and dignity, as in the old Firs that clothe the hills of California, Spain, and Switzerland. It is not in distortion or in little green bosses we look for the beauty of the Pine, but rather in storm-tossed heads and often naked stems; and all these ridiculous forms should be excluded from gardens and nursery catalogues. Another unfortunate tree, as helping to fill our gardens with useless and graceless things, is the western Arbor-vitae (*Thuja occidentalis*). This, which is a very hardy but never a dignified tree, even growing in the North about Lake Superior and through the Canadas, is unhappily also hardy in our gardens, where it forms a hedge or tree of some size, but has none of the qualities of the great Pines. Yet of this in one catalogue there are no less than twenty-three forms, all dignified with Latin names, and all absolutely useless for the true Pinetum; and though these varieties may have some value as knots of green in the garden, they are confusing when one wants to get the true Pine. There are many beautiful things, new and old, without filling our gardens with absurdities, many of them variegated; and of all the ugly things we know nothing is worse than the variegated Conifer, which usually perishes as soon as its variegated parts die, so that it often seems a bush full of wisps of hay. Moreover, these curious and monstrous forms of Pines are grafted for the most part, which makes them all the more useless to the planter who expects a return some day for his labour.

Pine Wood North of House.

The Holly as a Shelter.

In November and December of this year planted many Hollies round new kitchen garden and also in a fence round the rocks in Mill Place. The other day I saw some old Hollies in a high natural fence round a farm in Warwickshire, which well showed its qualities as a protector. The trees were old and the shoots feathered down to the very turf of the pasture, leaving sheltered corners all along the north-east side of the field, one of those home paddocks where young calves and cattle are often put out in Winter and Spring to enjoy the sun. In all localities exposed as most farms are, an immense difference may be made by making shelters of Holly. It is not, as some think, a slow grower if in a good soil protected from the creatures that love to bark it when small. Near pastures in woodland districts, or where stock raising is of more importance than arable cultivation, nothing can be of greater value for fencing or shelter than lines and branches of Holly not clipped, and the best way to make them is to begin with healthy seedling plants—never old stubby ones.

1893

The Holly as a Shelter.

Fences.

The men in the district making the fences too weak, we sought for a good fencer in Kent, and set him to work in Mill Place to fence the Pines, now an arable field. He makes a far better and stronger fence than our own men and uses much more material, binding it together too with a far stronger 'Heather'. The fencing on an estate like this is a cause of much expense and little satisfaction unless thoroughly done. In our own fences now we use strong Quick and a Larch pole along the top, as before described, so as to avoid the waste of wood and labour in farming the 'stake and Heather' fence which so soon rots and goes to pieces. Moreover, Quick does not grow well overshadowed and is far better well exposed to the sun.

1894

Cockspur Thorn. Having noticed this forming an effective fence in France, planted as many as I could get this year in the fence west of Moat Mead and also in the fence by Oak Grove. Bought also a quantity of seed in Italy, which I hope to raise so as to fully test the qualities of this thorn for fencing. I shall sow this seed in the field in Mill Place where the Oak was sown in Autumn, in alternate lines with the Oak, removing the Thorn after a few years' growth.

Planting Willow Sets. Planted March 1st a couple of thousand Willow sets of the Red Willow, which is used for tying in Covent Garden, along the stream in the Long Mead, and also all along the river down to the bridge across the road from East Grinstead to West Hoathly. At the same time planted a number of sets of the White Willow in the streamlet from the Moat to the end of the Crabtree Field.

Oak from Seed. Sowed with Acorns part of the Marl Pit Field near Birch Farm House—the Oak in the Wood west of the slip being very good and clean timber, I thought the same soil would give a good chance to seedling trees. Sowed the Acorns in lines six feet apart, sowing Cockspur thorns in lines between; these thorns to be removed as soon as strong enough to form fences. The Oak to be thinned early and regularly so as to give full room to the best trees; my wish is that the young trees should stand close but never so close as to weaken or injure each other; close and strong too is the way to get good Oak timber. The danger is in forgetting to thin the seedlings early and freely so as to prevent starvation in a mere thicket. (Sown in April, 1st week.)

Violets. Enjoyed these for many weeks this Spring, coming with the early Daffodils, in sheets forming a carpet to the shrub beds on banks north of house. Everybody loves the Violet, though few cultivate it in a way that is artistic or satisfactory. Many have not a soil to encourage its early growth, but should not be deterred by that. Violets come so abundantly from the South of France that in the markets they are not of much value, but that is no reason why we should not enjoy them much more than we do. Three years ago at Athens, in the King's garden, I was charmed with the masses of Violets in sheets over the beds. They scented the whole place, and were so full of flower that their effect told even at some

1894

Violets.

RISE OF MEDWAY, MILL PLACE

distance. A similar effect might be got in many English gardens; the little warm borders where we grow them for cutting are not nearly enough. One of the blots on our gardens is the bare ground among newly planted shrubs. An easy and artistic way to get rid of this is to cover the ground with some green thing which will carpet it until the shrubs come together. None is better than the hardy forms of the Sweet Violet. A good time for planting Violets is when they have done flowering, when they should be taken up, thrown into water, divided, and at once planted thickly, when they may remain two years at least in the same position, perhaps more. They would tend to keep the ground clean, and in a gentle Spring like the present would be a delightful feature in the garden; be useful for cutting, too, have a good effect in the garden landscape,

1894 and scenting the air at the time of the Crocus and early Narcissus.

Flowering Shrubs in Fences. At the end of March sowed many seeds of shrubs in the newly formed fence banks by the Oak Grove, and also surrounding the Rocks in Mill Place. Among these shrubs were the Almond, Medlar, Wild Roses, several Plums, Greengage, Mirabelle Prune, Bird Cherry, also climbing shrubs like the wild kinds of Clematis, i.e. *C. crispa, C. Flammula, C. Viticella, ligusticifolia, montana, orientalis,* and others.

Stout Herbaceous Plants in Hedgerows. Taking advantage of the newly formed and mended hedge banks resolved to sow or plant in them a number of stout herbaceous plants. These plants, often a short time in bloom, and therefore requiring much space in the garden, hardly deserve it when we think of the long time they are flowerless and the large gaps they leave after blooming. Naturally they grow in woods, copses, rich grass, and in situations by no means always so favourable as cool deep hedge banks, so I thought we might enjoy their bloom quite as well in these and free the garden for other things, and, as the readiest way of establishing the herbaceous plants, sowed seeds at the end of March, among them being: Anchusa, Everlasting Peas, Rose-Campion, Ferula, Lupins, Heracleums, Delphiniums, Columbines, Aconites, Bellflowers (tall), Cotton-Thistles, Giant Poppies, Telekia, Goat's Rue, Baptisia, Giant Scabious.

Almonds for bloom. Planted many Almonds on the Rocks, seedling trees, and also sowed many seeds of the Almond in hedgerows at the end of March. In March the Almond tree is often a picture in many London gardens, especially in the southern suburbs. The question worth asking is: Do we make artistic use of this very early blooming tree, so bright and delicate in colour, and so effective in the landscape when healthy and well-placed? We have many flowering trees, but none of the quality of this, that comes before the swallow dares, and earlier than many of the hardy Narcissus. No tree lends itself so readily to the making of living pictures; but it is usually so primly set in the garden that we do not get half its value. The grafted stem and often round head are the rule, but by no means are they essential to the Almond. One I saw this year was prettier far. The tree had broken down when young, and, not being raised, shot up a few long shoots in its own way, and was a beautiful and paintable object. It would be very easy, by planting Almonds in groups and masses, to get a happy change, not in prominent places, but in rather poor half-waste spots of no great consequence in the garden or pleasure ground. They would go well with our Crabs and Sloes, and would be quite as easily raised. There are heavy soils in which the Almond does poorly, but even on such it is often possible to find a suitable corner for the tree. In the Thames Valley Almonds are very happy, no doubt by reason of the deep, and at the same time, warm and good soil. If people who are confined to a small patch of garden in a London suburb can yet make a picture of their Almonds, those who have country places should do very much more with so fair a tree. It is far too good a thing to be content with one or two trees of it, and yet does not bloom long enough to be worthy of the best positions. To use it with rather a free hand, on banks and in places of little consequence, is the true way, and for those who are in their gardens in Spring nothing will give a better result. 1894 *Almonds for bloom.*

Between the farm and hedge banks and rough slopes, sowed this Spring as much as $2\frac{1}{2}$ cwt. of the seed of the common Furze, sowing the whole of the Kingscote Field to test the plant for feeding stock. The beauty of the plant in March this year was charming in many positions, and the field, sown a few years ago, had been very useful to us during the Winter in the farmyard. In a large part of Britain the best covert plant is the common Furze, and, to a less extent, one of its allies, the dwarf Autumn-flowering Furze. While so many are making experiments with exotic shrubs, it is well to keep in mind this *Furze.*

1894

Furze.

excellent bush which, in our mild climate, perhaps attains its northern limit, not being a northern plant really, as is proved by its dying in very hard weather, and in France keeping much to the sea. In our climate it grows a good deal more inland, and over a vast area comes of itself. The great point in favour of the common Furze is quickness of growth in places where it is not already common. Naturally it grows on warm, sandy, and poor soils, but will grow on heavy and cold soils. No other covert plant can be so easily raised. A few years ago I sowed a heavy field, with a northern aspect, which was hard to work. The plant came well the first year, and the second there was a very fine luxurious crop, and so it has remained, and a grand covert it makes. We did not wire, although there were many rabbits about—the plants came so thick and so many that the inroads of rabbits did not hurt them. The beauty of the Furze should be of some account in our country, as it so often blooms early when there are so few other things, and its effect upon the landscape is good. Apart from its use as a covert it is of the greatest use on the farm, and for the shepherd to make his little lamb-pens and yards nothing is quite so good. In heavy ground it is excellent for bottoming stock-yards and protections of various kinds, and also for roofing rough shelters. In Winter when fodder is scarce the crushed shoots are good food for horses and cattle; and in certain soils it is worth more attention from this point of view. In poor lands, as in Brittany and in other parts of France, it is grown as an important fodder crop. We have two kinds in bloom—the common and the Foxbrush Furze, the latter a handsome one, of which the shoots form a long plume, sometimes quite set with flowers. This Furze is very little known in England. The Dwarf Furze, which abounds on many English heaths, and so much in Ireland and the west-country, is not so large as either of these, but is a very pretty bush. All these kinds can be raised easily from seed, but the cheapest seed is that of the common Furze. We have raised all the three kinds of Furze at Gravetye by sowing the seeds directly where they are to grow—the simplest and the most satisfactory way, though Furze will grow freely when transplanted in the young state. In no other way could we have raised so quickly seven acres of dense covert at any price. Whether from a covert, an artistic, or a farming point of view, there are few plants so important as the common Furze, Gorse, or Whin, which is a native of Europe, from Denmark to Italy, chiefly in the west of Europe, and in our islands. 1894 — *Furze.*

Laburnum. Among the many other seeds of shrubs and bold flowers sown to take their chance in hedgerows, banks, and in rough places this Spring, sowed Laburnum seed on the railway banks near Birch Farm, having noticed how pretty little seedling bushes of Laburnum are abroad. Sowed also in various places about the estate White Broom and Spanish Broom.

Royal Fern. Found, May 18, to my surprise, a few plants of the Royal Fern (*Osmunda*) in a moist part of the recently-grubbed shaw on the lower side of the Crabtree Field, probably the last locality in the place for this noble native Fern once wild in the district, from which it has been exterminated by London hawkers. A few little seedling plants probably escaped in a corner of this remote shaw.

Jarretts and Hoathly Fields. These two fields being narrow, small, and awkward to work, and much infested by rabbits, removed the shaw lying between, and formed one large field for arable work with an easy woodland boundary. The effect was very good, and with thirteen acres or so of ground it will be much better worth looking after. The field will for the future bear the name Jarretts, the old name of the larger of the two fields thrown together. June 4.

Collection of Tulips. Formed a collection of Tulips this Autumn, planting them in long borders through the orchard each side of a grass walk. Over them, and to succeed them in bloom, we planted many of the choicest kinds of hardy herbaceous flowers, the whole

1894 backed up by a trellis which is to be covered with Roses and the more graceful climbers. The Tulips consist of the wild species and the more beautiful and distinct single varieties. September and October.

Collection of Tulips.

Sussex Cattle.

Added many fine cows to the herd in Autumn from sales at Chellowes Park at Mr. Woods (Hassocks), Ashenden, Ivychurch, Robertsbridge, and Maybanks: buying the finest young cows and best heifers we could see.

OLD ROAD TO HOUSE

1895

Bamboo Garden. Being much interested in the Bamboos, and the garden soil and situations in the garden itself not being very favourable to them, commenced the formation of a Bamboo garden in a little hollow with rich good soil in Warren's Wood near the new stables, intending to approach it by one of the woodland paths, and keep it as secluded as may be. We do not propose to add any soil or preparation beyond grubbing the stumps of the underwood. The site looked very pretty on being cleared, and we propose to plant in April all the hardy kinds that can be got, putting each kind by itself and associating with the Bamboos Japanese herbaceous plants and Azaleas.

Reservoir for Water Supply of House. This having shown signs of leakage, had it examined and found it very imperfectly cemented against walls of the local stone which have no value for such work. The walls were so bad that the water leaked out in places and did not rise to half the height it should have done. Emptied the reservoir, lined it with 9-inch brick well grouted with the finest cement and Thames sand, and then cemented it all over with the same cement. The archwork was rough and open, showing the light in places; put a strong mass of concrete over this and cemented the inside so as to make a permanent thing of this reservoir. Carried the brickwork up so as to give some additional support to the arch, and made the whole of the inside as near as possible like a shell. Put iron doors and one ventilator; threw back the soil when the arch was concreted and planted Ivy on it, intending not to plant any trees on or near the tank.

The Sussex Sale. After seven or eight years' experience of Sussex cattle I find them hardy useful cattle, and perhaps the best that could be got for the soil, as they live where others would starve, the grass and provender of the country generally not being good

1895

The Sussex Sale.

for the more delicate and choicer breeds. They may have certain drawbacks—the increase is not always so rapid as one would like, and barren cows occur, but they make excellent beef and better than any breed tried in that country. About their beauty in the fields there can be no doubt; they are a beautiful dark red colour, and the manner in which the cows, calves and all group themselves in the fields is a constant source of pleasure even in mid-Winter and perhaps best in the Autumn. Having kept Sussex for a number of years they became too numerous for our land and food, and so I determined to sell the calves of the past two years and the bullocks, but the auctioneer persuaded me that I should have a better sale by selling the whole of the pedigree herd, which we accordingly did, omitting the non-pedigree animals. The sale took place under good conditions and was most carefully managed, and the result, as shown in the following pages, was over £2,000, the calves and bullocks selling well and the cows below their value; but on the whole the full value of the animals was probably obtained.

As to profit from such a pedigree herd, under present conditions it is out of the question, and it was carried on at considerable loss every year. It is probable that this loss was increased in the desire to have pedigree animals which were often bought at a high price. Unless there is a real demand for stock animals (which can rarely occur save in the case of well-established dealers), it is probable quite as good results could be got from non-pedigree animals, as the pedigree makes no difference to the butcher, and in fact the difference in the price of pedigree animals and others is sufficient to account for the absence of profit. There may be other reasons also. With farming as it is at present it is hopeless to expect any better result. It is probable that breeding Sussex cattle here of the ordinary stamp would lead to as little loss as any other use of the land, but loss must be faced in any case. I was sorry to

The Sussex Sale.

see the fine cows and heifers so below their true value, but on the whole I could not say the sale was not a success. There is such an immense loss, manage as one may, on any kind of farming, that the only safe way is no doubt to let the land. Then at the worst we know where we are, whereas with amateur farming and all its drawbacks the loss is great and the trouble unending. It is fool-farming, and one has not even the pleasure (on my land) of seeing a good crop in return.

Tea Roses in 1895.

This year, owing to the fine Summer and better preparation of the ground in previous Winter, the Tea Roses did remarkably well. Many people thought that the great severity of the Winter would have killed our Tea Roses. At first they were cut to the ground and looked very weak at starting, but the steady growth throughout the Summer was succeeded by vigorous growth and splendid bloom in the Autumn, and in all the beds, which had been deeply prepared in previous years, the result was beautiful to see. By this time, after many trials, we had found out the best kind that open in England, and these in bold masses did admirably. Their colour was at the end of September more 'showy' in mass even than the 'bedding out' in the best London parks, while infinitely more beautiful seen at hand. The fine bloom continued to the last day of September and was indeed charming on the first day of October: on the night of that day a heavy cold rainstorm put an end to the fine show for the year.

Artists in the Garden.

Four different landscape painters painted our Water-lily pond in the Summer and Autumn of 1895, viz. Mark Fisher, Alfred Parsons, H. A. Olivier, and H. G. Moon. Mr. Fisher found plenty of other pictures about the place and painted some of them. This was gratifying to me as proving that the garden can be made artistic and beautiful to those best fitted to judge. They said the place and garden and all was full of pictures. I have worked long to prove that the garden, instead of being a horror to the artist, may be the very heart of his work. There

1895

Artists in the Garden. is no reason why the garden, which in our country is so often the foreground to a beautiful landscape, should not itself be a picture always.

Work Winter and early Spring, 1895–6

Kingscote Field. Commenced to plant with the hardiest Pines of Europe and others equally hardy likely to form evergreen forest trees. February.

Kitchen Garden. Moved this from the Highfield to the lower part of Warren's Field, where it was originally made by me, as we found there was no water on the hill-top. December and January.

Quick Fences. Fenced the whole of Jarrett's field with strong Quick, using chiefly old Quick from own woods with a good many stout Quick bought in nurseries between. Fenced also the Sands Field in same way. The whole of the fencing was begun in early Autumn and completed by the middle of March.

1896

Young Larch Woods.

Threw out all the young trees affected with Larch disease from the Marlpit Wood, and as there were a good many it amounted to a thinning of the Wood. Dr. Schlich, who visited the woods with me, thought we might fight against the results of the disease in this way; but probably, mixed planting is the best remedy, if there be any.

Fish in Large Pond.

Drained and took out all the carp, which were always a nuisance here. Found very few trout owing no doubt to the presence of the carp. Owing to leakage in the wall of water-shoot, decided to do away with the shoot and revert to the old way of getting the water over by a horseshoe embankment. The wall removed, laid bare the old foundations of a circular embankment, which we rebuilt in the best way we could and hope it will endure.

North Bank above House.

Changed this much by removing the Yew trees and securing more airy effects about here. Formed large beds of Summer-leafing shrubs on the bank which had been rather bare; the shrubs grouped in picturesque masses. Planted many Heaths here and Rhododendrons on their own roots.

Boat-house.

Filled in narrow neck between shore and Island above boat-house to allow of some Willows being planted there to hide a little the end of the boat-house.

Water-lilies.

Planted various kinds this Spring, mostly received from Mons. Latour-Marliac and some from North America from Mr. Gerard.

Kingscote Field.

Planted this with hardy Pines of Europe, beginning in January and hoping to complete in Autumn. Thought it would be better for game to have a wood of Evergreen trees here, only planting those of proved hardiness.

Yew Trees in West Garden.

The group of Yew trees having a rather lumpy appearance, took down the smallest of the trees which had almost no top, in order to give space and better form to the remaining three; the effect is better, more of the stems of the trees being seen. The difficulty will be to keep the roots out of the flower beds.

1896

Iris Garden. Formed in the Summer this year two great Iris borders, with grass walk between, to be the Iris garden. Planted in Summer and Autumn the German Iris and any other kinds likely to thrive in a not very rich soil, which was first lightly trenched and manured. Having planted the Iris, some biennial and Spring-blooming plants, like Wood Forget-me-not and Iceland Poppy, were sown through them, in order to give an earlier bloom and help to furnish the border till the Iris get strong.

Quarry. Worked the quarry in Beeches for Mill Place Farm work, and found the stone good in texture and excellent in colour; used it in the new garden retaining-walls and in the new wall to Mill Place Farm.

Bridge in Brookbank's Field. Built in the Autumn this year a strong two-arched bridge of brick and cement across the stream in Brookbank's Field. All attempts to make bridges over streams of wood, however good, having failed, resolved in future to build all such bridges of brick and stone tops. Wooden bridges over streams liable to spates often come to grief in a few years, and making such bridges is a great waste of labour and timber.

SANDSTONE ROCK: MILL PLACE FARM

1897-8

Walls in West Flower Garden.

Took down a large Redwood tree on bank on west side (on bank outside garden) in order to complete the group of Hollies at the end of this garden, and so get all about the garden the effect of native trees. Did away with the slopes border round the flower garden at the N. and W. sides and built strong, dry walls of our own sandstone. Each stone was laid on a line of Alpine or rock plants with the merest pinch of soil or sand under the plant. These retaining-walls round the garden will enable us to have level borders instead of the sloping ones which starved in dry seasons, and will be in other ways a gain. As the wall was built up it was backed very firmly with broken stone, sand, and good soil, so that the rock plants will find plenty of nourishment.

Narcissus.

Planted in January nearly 100,000 Narcissus, mostly along the margin of the Lower Lake and in and by the protecting fence of same. The bulbs were planted late and, perhaps, slightly the worse; but I hope in the end they will take very well.

Sweet-brier Covert.

Continued this round Lower Lake from where it stopped half-way, thus giving easy access round the lake to the dam. The soil was of little use for grazing, and the covert brought it, as it were, into the pleasure ground, and it will be a gain in various ways. Planted Cockspur Thorn and Sweetbrier in the fence bank, and many trees and shrubs in the covert—mostly trees of a dwarf or medium stature. Hollies and Yews being planted towards end of lake both for look and shelter.

Magnolia Garden.

Planted the piece of ground first intended for Bamboos with Magnolias of all obtainable kinds, as the Bamboos seemed not to thrive; and I thought better to plant the place with

1897 these handsome trees, which in most cases are likely to be hardier and give beautiful effects.

Evergreens about East Front of House. Owing to the movement of soil when the surface was changed by me here at first, things have not done very well here. The site is not well fitted for an orchard, and I resolved henceforward to depend more on Evergreens and plant most of the East front and Approach side of House with the hardiest of these. The bank above being already planted with Holly and Yews and the Scotch Fir groups about the House, all native Evergreens, the difficulty was to find Evergreens from other lands as hardy. Planted Kalmias, Savin, larger varieties of Heather, Alpine Rose, Golden Yew, the hardiest Berberries, and the large Partridge Berry (*Gaultheria Shallon*), and Rhododendrons of the hardy American sorts on their own roots. Also *Eleagnus*, *Osmanthus*, *Skimmia*, *Andromeda* (or *Pieris*), and hope to get some native Juniper; but it is not much grown in nurseries, though Junipers from almost every other country are to be had in them.

Walls by Pathway to Stable from House. Rebuilt these retaining-walls in the Spring this year. Eight or nine years ago, when we made the way to the stable through the bank, they were very roughly done and packed up with turf, notwithstanding which they gave some pleasure almost every year—the rock plants flowering and thriving well in the rough walls. Encouraged by this we rebuilt them with greater care, using a more regular stone and no loose soil or turf, but putting small plants between the stones with the smallest dusting of earth between, packing the stones behind with good soil, sure that the roots would soon find their way to the back, all stone being set in rock plants of some beauty, or small native Ferns. Completed in early Summer, these walls were in flower throughout the Summer, and in the following Spring were a sheet of lovely Alpine and rock plants.

Trout in Lower Lake. On March 25th put in the Lower Lake 500 trout—450 six-inch yearling Fario, 25 ditto Levens, and 25 ditto Rainbows

—costing, with carriage, £14 19*s.* 6*d.*, and bought from the Surrey Trout Farm at Critchmere, Haslemere. 1897

Earth-slips. These were rather frequent this year, owing to the very heavy rains; and in some cases we suffered great danger by the slipping down on roads, as in the Rocks Field, by the road between West Hoathly and East Grinstead. Many hundred tons of earth had to be carted away, and eventually we had to have recourse to deep stone drains, getting below the moving ground with the drain. Blasted some stone in the rocks above and put them in here to form the stone drains. The landslip on the drive was also troublesome, and was dealt with in the same way. After many fillings in of earth and gravel, we had to go back to the deep stone drain.

Cider Orchard above Home Farm. Planted in January and February a collection of the best cider apples known to French growers. Some of them, very fine examples of well-grown nursery stock, grafted low, so that the stems are formed of the Variety, and not the Crab on which it is grafted. Some of the newer kinds, however, were grafted high on the stem of the Crab in the old and, I think, worse way. All were planted in the paddock above the Home Farm yard and north of it, laying their roots on the surface of the earth and raising a mound of road scrapings, or any other loose earth at hand, over them so as to keep the roots a little out of the cold, sour bank, of which a good deal of the North paddock is made. A deep hole was first dug out, driving a stout stake in to support the tree, using some old Oak rafters from a demolished stable. Strongly staked, the trees were then wired to keep off rabbits and cows, and an inch or two of loose stable manure thrown over the mounds to keep them moist. In planting, a few spare roots of Narcissus were thrown into the mounds in the hope of adding to the beauty of the spot with Spring flowers. We were encouraged to do this by the fact that some varieties of the Star Narcissus put in in the same place eleven years ago had thriven and flowered beautifully

1897 every year without any attention and without any noticeable lessening of the size of the flowers, or any of the disease showing itself, of which so much is heard as to Daffodils grown in cultivated ground.

Lower Lake. More than half emptied and a good deal of the bottom cleaned, and the mud put round the edges so as to encourage a strong growth of herbs around the water for the sake of the fish as much as for effect, until a rush of water in April made us give it up for the year. Began to put some of the newer Water-lilies round the bank this year.

Narcissus. Planted about 80,000 Narcissus mostly round the Lower Lake on the fence banks and by the margins of the water in the most hurried way, putting them in cattle tracks and hollows and covering them with mud out of the lake.

Sale of Farming Stock. After a good many years' experience and continual heavy loss, resolved to give up every kind of farming on my own account and let the farms. This is perhaps the necessary consequence of taking up a business and not giving it personal attention, which in my own case would not have been skilled attention. Farming is often supposed to be a simple matter, but is really a complex business demanding much good and early training and varied experience, natural ability, and love for the work. It was always so, and in these hard times for farmers it is very much so, and any one who undertakes to farm without knowing all about it, and what can be done in the district he works in, is sure to fail.

Pergolas near House. Formed pergola with massive stone pillars as entrance from the east side of the House to the garden, planting Wistarias on it. Formed also pergola from stable to kitchen using Bamboos, stout Oak supports.

Little Flower Garden. Edged with stone all the beds here and renewed some stone paths that had been poorly laid. Refilled and gave a complete change of soil to most of the beds here, planting more Tea Roses.

UPPER LAKE: DAFFODIL TIME

1897

Bank North of House.

Altered wholly the plan here, giving up the path beneath the Scotch Firs, and beneath the great Oak, planted a broad belt of Yews and Hollies to cut off the view of the road and new kitchen garden here, and formed a grass walk below this, which gave a better view of the House and its surroundings. Planted Poplars temporarily among the Yews to give a better screen, intending to remove them in a few years, when the Evergreens have taken hold of the ground. Planted a great number of handsome well-grown seedling Hollies here this May, 8 to 10 feet high, mostly brought from Berkhamstead in Herts.

Stoned West Garden Walks.

Resolved to do away with the gravel walks here and pave the walk round the garden with stone, using old London York stone used for paving, except the part under the Yew trees which was filled with roots, and was too irregular for paving. The stones were laid at 'random' when they did not come in well any other way. Laid by John Fell in sand. Formed also a paved bed here on which to place plants in tubs out for the Summer, if we can get a good set of fragrant plants for this purpose (plants that cannot be grown out of doors in the district). In the crevices of the stone-paving planted very dwarf rock plants: Thymes, the Fairy Mint, Creeping Speedwell, Tufted Harebell, and like plants. Stone pathways in this and the little front garden completed last week in July. The front garden, which was done several years ago, had to be in part relaid owing to the use of too small pieces.

Sowing Trees direct.

Sowed Coneyburrow Field in Mill Place with the seeds of some of the forest Pines of Europe: Spruce, Scotch, Austrian, Corsican, Silver Fir, Larch, and the Canadian White Pine, with the common Juniper for covert, as an experiment, and to compare with the results of planting trees as usually carried out. It was a fallow field, and was harrowed well before sowing and once afterwards. One often sees trees such as the Scotch Fir and Larch 'sowing themselves' in the worst conditions, and

1898 — *Sowing Trees direct.* good, clean seed from the best source sown on fallow ground should give a good result. The seeds were sown broadcast. It might be well to sow in lines 4 feet apart, as, the seeds of the Pines being rather heavy, it is not so easy to get a fair distribution of them over the ground; but if one of the disadvantages of sowing is that the seeds may come thicker in one part than another, transplanting the seedlings in the field is much easier than if we have to bring trees from distant places. My reasons for sowing were various; good plants for forest planting are not always easy to get, and there is a way of selling them too large, which is often a cause of great loss to the planter. As the ordinary nursery does not produce forest plants in just the right state for woodland planting, nurserymen have very often to apply to others for them, and hence a double movement of the plants, which in a dry Spring often suffer injury in coming long distances by rail. Unless also the ground and men are ready on the day of their coming, the plants sometimes suffer after getting into the place. The best results are from little trees, say under a foot high generally; but during hot years, in the Southern and Midland Counties, failures are common even of sound little trees, and though the plants may not be entirely lost, there is very short growth in some years. Planting as usually done is expensive, the amount of labour involved in planting a few acres being considerable. This is especially so if the work is not controlled by a forester who knows what planting should cost and how quickly it ought to be done. An active man, on the other hand, can sow ten acres of seed a day. I shall have to wait long for results, and objections on the other side are frequent; but every art has its routine ways, which when looked into sometimes are not always the best ways for the purchaser, and are often nothing more than a trade convenience. In any case the raising of a wood from seed is a very interesting way, and, being also the most natural, is well worthy of a trial. Where a mixed woodland is desired, it is very easy, by mixing two kinds of seed not very dissimilar, to get good associates such as Scotch and Austrian Firs, with a sprinkling of Ash or any other hard-wooded tree that may be preferred. I sowed broadcast and thinly: but if I had sown in lines it would be easier to give the little trees a better chance by using between the rows a light hoe and keeping them free from weeds, and also thinning would, perhaps, be a little easier; but seeing the way Larch and Pines often sow themselves in the most unlikely places, and often on shaly banks without soil, I resolved to chance letting them fight it out on the grass, the only care given being to wire against rabbits. In this way, also, the field would be a better covert for birds, which with many is a consideration.

1898

Sowing Trees direct.

A month later (May 28), sowed in the same field a few scattered seeds of the Atlantic and Lebanon Cedars sent me by Mr. Posth, of MM. Vilmorin, of Paris, just putting them in little hollows and pressing in with the foot. I propose to call this field (when it starts into a wood) Coneyburrow Wood. These Cedars I believe to be the same species or best forms of the same Cedar of Lebanon.

Water-way from Upper to Lower Lake.

The old waterway here having become broken up owing to bad construction, I formed a new one, strongly made, using natural stone and good concrete backing for the walls of the same. The material used was 20 tons of sand, 10 tons of burnt ballast, about 10 tons of cement, and 2,500 bricks and much natural stone—in all about 100 tons weight. Finished May 27. The object of making this, apart from the passing through of the water, is to get a more picturesque effect, with a view of the Moat Cottage in the distance, and also to help in the formation of a Bamboo and large grass garden here, where Bamboos and bold grasses may be seen in association with Willows. Care was taken to avoid any appearance of 'rock work', in the ordinary sense, by using no more stone than was necessary.

1898

Bamboo Garden (New).

Having filled in the old waterway from the top to the lower gate, levelled the ground and commenced the plantation of Bamboos on broken ground, planting plenty of the common Japanese kinds and two or three others, the object here being to associate the hardier Bamboos with Willows in such quantities as will allow of their effect being well seen. They took well the first year, planted in early Summer. Some taken up in November were quite strong at the root, and had even pushed out stout stolons. They were well 'mulched' all the Summer with any waste materials near—weeds, stems of flowers, and quantities of the green scum pulled off the lake. This had to be removed several times during the Summer, and all of it was placed over the surface below the Bamboos, so that they never suffered from the drought.

Smugglers' Lane Walk.

Opened up this in Autumn, bringing the walk through the Iris garden right into it: the shade of the Oaks will be grateful in hot days, and we shall have a picturesque way to the Lily pond. The form as well as the shade of the Oaks will be good here. A view into Moat Mead and other fields near very pretty when the lane was well opened up.

Heath Garden above House.

Formed a garden by the Ilex Walk above the playground of the hardy Heaths of Britain and Europe in very bold masses. I believe this will take care of itself with a little aid at first, and be a very pretty and interesting garden. The beautiful native and European Heaths, which are among the most delightful of all small shrubs in their native conditions, are very much neglected in gardens, and even when grown are rarely or never grouped in any effective or picturesque ways.

Stone Wall round Fruit Garden.

Commenced to build in early Summer a stone wall round new kitchen garden, chosen after much and long thought as to the site. For various reasons chose the open hill above the House (where the spring that supplies House arises) as the best, and indeed only good, site near the House for the kitchen garden. Got out own stone at the 'Beeches' on the estate, said to be

1898

Stone Wall round Fruit Garden.

the hardest and best of the stone in the district. Wall built by the men who got out the stone. I prefer stone to brick both for its colour, endurance, and the fewer interstices for harbouring insects. An oval plan was followed to suit the ground: as the usual rectangular plan would have led us into awkward angles and levels. Good foundations of cement, concrete, and rough stone above it were put in. Planted first Peaches on the inside of the wall so far as it was built in late Autumn. Foundation-stone laid July 28, 1898; not completed until middle of July, 1900.

Yew Trees SE. of House.

Planted these round new outdoor closets to conceal them from view. The way the Yew breaks into many stems mars it as a young tree. I do not think this is quite natural to it, and is probably the result of using cutting plants instead of seedlings, the naturally raised trees being the best always.

1899

Covered-way from Kitchen Door to Stable. Formed the main part of this in early Spring, using own Oak supports and Bamboo stems for the lighter work, and planted mostly with Vines and sundry Creepers. Planted and formed the little dry stone wall at the same time.

Rhododendrons on own Roots. This year began to plant some of these on own roots on the beds (large) on the E. side of the house. Plants from Dickson's Nursery at Chester, very good ones, and from a loamy (not peaty) soil. In this way hope to get over the usual result of growing all kinds of Rhododendrons on the Pontic kind, which mainly proves that the hardier forms may be grown in brown soil in a better way than is the case when they are grafted.

Stone Stairway in Smugglers' Lane. Formed this leading from new orchard garden into the bed of lane, planting the supporting walls with rock flowers. Pergola to cover same made afterwards.

Tank in West Garden. Built this for the growth of the newer and rarer Water-lilies which we have often lost in the large ponds owing to the attacks of rats, waterfowl, and other enemies.

Larch Seed sown in Woods. Sowed broadcast and thinly in the woods, recently cleared of underwood, good fresh Larch seeds: Hastings Wood, Stockards, part of Warren's and Bushy Woods being sprinkled over in this way for the chance of occasional Larch coming here and there in the woods. Larch growing here and there in mixed wood is often remarkably good. As my aim is to reduce the underwood, now almost valueless, and encourage as much as possible timber growth, it seemed to me that the seeds of Larch might help in the general changes, and that though a 'crop' is not to be looked for under any such circumstances, in bare spots some trees might come up, as Larch does now and then sow itself.

1899

Broom in Hastings Wood. Sowed about 6 lb. of the seed of this in Hastings Wood (this being the year of clearing underwood), sowing near the drives and by the brow along streamlet; if it escapes rabbits it should form a pretty feature in a few years' time. Sown in June.

Wall E. of House. Built wall here to shelter the garden from the N. and E., and also from the drive; made small Oak and shingle shed at end of it to protect machines, and on near side a Vine-clad bower. Trellised wall with Oak trellis-work, in the old way (instead of wiring). Wall about 120 feet long, built of local stock bricks, the coping from Milbank Prison.

Lower Drive to Station through Bushy Wood. Formed a road from House to station through Bushy Wood, taking line of easiest grade we could find on the hilly ground. Got a very good line; did not stone the bed of road for more than a year after forming. Took two years in all to form, thus completing the main roads into the place which were formed by me. I should have preferred one drive in, as disliking a multitude of drives, but it was essential to make a road to the railway station; and hence this drive. The gradation was carefully studied to avoid the sharp pitches, which are so disagreeable in many places. The absence of a good road into a country place is often a great drawback, and Gravetye was almost impossible of entrance for carriages before these roads were made.

A. BRAZIER: WOODMAN

1900–1

Summer-house in SW. corner of West Garden. Built this from design by Mr. Ernest George. Low sides of own stone, stone floor, sides of Oak framing, roof of shingle: all in Oak.

Pergola back of Stable. Built this with pillars of stock brick, oak plating, and larch cross-beams, with a view of covering the stable with vigorous climbers in Autumn. Planted Clematis and Japanese Wistaria.

Furze and Broom. Sowed much of both in the woods in Spring, May to June, sowing in the woods where the underwood was cut in Autumn—Warren's, Jarrett's, and also in fences or any bare places.

Sitka Spruce. Planted colony of this in higher corner of the Bushy Field, and fenced this off from field and also from rabbits: common Spruce planted between to be cut away when the Sitka Spruce starts into strong growth. Sowed also in same small field Furze and Broom, to give a little colour and varied 'cover' to the corner until the tall trees grow up.

Planting south end of Paddock. Planted the awkward and almost useless slopes and wet corners here with Silver and Russian Fir and Northern Spruce; fenced the corners from paddock and also against rabbits. Planted in February. Sowed also Broom and Furze among the newly-set trees and on the newly-formed fence to cut off paddock here.

Quick Fences. Formed much bank-fencing early in the year and planted many thousand Quick: planted with these the banks along new road being formed to station, also at bottom of paddock: along North Field, cutting it off from the Sands.

Stone Roof of Old House. The Horsham stone roof being in a bad state, took it off, and renewed it in Autumn of this year. Brought men for the purpose from Collyweston in Northamptonshire, where the art of stone-roofing is well understood. Began in October; Autumn mild, though very wet; hope to escape the effects of frost. For the smaller parts of the roof (gables) some of the thinner Collyweston stone was brought; but for the old roof in its main parts we found old stone in the district. Used instead of the tile-ridging Yorkshire stone-ridging from near Leeds. Put a sheet of zinc under the stone over the whole roof to keep out cold wind, bats, and birds. Renewed or replaced the very old stone coping, very loose and decayed in many parts. Put up lead pipes in lieu of iron ones; renewed lead guttering for roofs.*

1900

Stone Roof of Old House.

Tea Roses on Own Roots.

Commenced to reform the beds in the West Garden, throwing out the heavy, rich, used-up soil and replacing with free, light soil. Changed shape of beds at same time, doing away with margin of grass between walk and beds.

Rhodos on Own Roots in Woods.

Commenced to put these in Warren's Wood, in the hope that the layered plants will thrive in the brown soil of the wood and some day form an evergreen undergrowth.

Renewing Woods.

Began an attempt to get rid of underwood in Warren's Wood, planting tall saplings of Maples, Sycamore, Oak, Acacia, and Poplar, between the stubs, in the hope of these killing the underwood as soon as they get ahead. So near the House (even if not in every case) it is a mistake to carry on the old underwood system, with its ceaseless injury to the surface of the drives and roads and other drawbacks, while it has ceased to return any profit worth speaking of.

Rides through Woods.

Began to consider and change these into more airy and roomy ways, watching the line of each for its beauty as well as convenience for the shooters and any light carting through the

---

* NOTE: July 11, 1901. *Continued same stone roofing over all new parts, working all the Spring, finishing July 6, 1901. All the roof covered with zinc below the stone: old roofing stone found near Horsham and Charlwood.*

1900 — *Rides through Woods.* woods. Began in the Pine wood above the House, going thence in Warren's Wood (above drive) and connecting the whole so that the drives meet easily and give convenient access, as well as good views. Much of this was done to make the access from one wood to the other easy without gates, and no work ever gave me more pleasure or greater satisfaction.

1901 — *Stone Roof of Manor House.* This work completed July 6, having taken about nine months' work in all.

Pergola round West Garden. Formed this in Autumn of Oak and Larch, with paved path beneath to run round west and north sides and give a little shade in hot days. Planted with Wistarias, Roses, and Vines; six plants of the Vine classed as Royal Blanc (said to be an early and good Vine) were planted on pergola near the House.

Oak Paling above House. Formed this year to give protection to Pine walk, planting having been tried and found useless to cut off view from road over ground here, owing to Pine trees overhead. Planted on the inner side with various kinds of Honeysuckle. Fence made by Mr. John Longley, of Three Bridges, at a cost of about 10*d.* a yard.

1902

Replanting Woods.

Early in year (January) replanted some of the woods, in the hope of getting rid of the underwood with its various drawbacks. Oak, Ash, Beech, Canadian and Black Italian Poplars, Wild Cherry, White Willow, Norway Maple, Sycamore, and Abele, the trees mainly used. These were mostly tall, fine saplings, often 12 ft. high, brought from nurseries at Oudenbosch, Holland, where trees are well grown for such planting. Among the woods replanted in this way were Mine Pit Wood, which was very thin of trees, only a few poor scrub Oak being left—this absence of trees of any value being partly the result of the underwood system, a profitable one at one time; Warren's Wood, lower part and bottom; Bushy Wood, bottom, near culvert in new drive; Railway Shaw, i.e. slips over railway (N. side); Drive Shaw, mainly with Willows and Canoe Birch; 'Alders' Wood, sloping into Valley from Upper Drive, mainly Oak; 'Pit-cup,' a slip below same drive near outlet; Shaw near Moat west of 'Larches'.

Upper Lake.

Having got this silted up, removed about 15,000 yards of mud from it, and used a good deal of this to enrich the banks of the pond; but the main part was put over the South Lawn above the pond, as I thought it better to get rid of the mud at once rather than turn it over and dress with lime, as is usually done.

Bowling Green North of House.

This, which has been an immense trouble to form, was turfed over, levelled, pressed firm, and again returfed this Spring. Some inches of rather sandy soil from the drives in Warren's Wood were first added to it in the hope of making the turf better and more able to withstand dry seasons. This is the third attempt in fifteen years to make this a good turf; but, owing to the fact that it was originally laid on partly-made

1902 ground, sinkages and other defects have been observed. It is now being laid by James Burnham, who is the best ground workman I ever knew for forming lawns of this kind. Turf brought from Vowel's Field.

West Garden. Made a great change here by taking away the whole of the turf of the central lawn and throwing the space so bared into simple square stone-edged beds with stone paths between. The turf, I thought, was not needed here, especially with the play-ground above the House quite near, and the ample area of grass on the south lawn and elsewhere. I also thought it better to concentrate the interest on the flower garden, and turf is not essential for small spaces near the House.

Magnolias. Made another attempt to grow these fine shrubs and trees, and opened up the continuation of the Smugglers' Lane as it went up the hill. It was underwood when I came, but, having planted Corsican Pines, Cedars, and other Evergreen trees, it now forms a sheltered little vale in which I thought the Magnolias might thrive better. So we filled the hollow with good soil from the hill above, made a turf walk, and planted a fine series of Magnolias, of nearly all hardy kinds, from Leyden—handsome plants, full of flower-buds when they came. Planted in January. Flowered freely in April of same year.

Lower Lake. Lowered and cleared of mud the early part of this year; that is to say, all the shallow parts where the weeds grew, not the central parts where there was little mud. Added some new groups of Water-lilies: cleared away Ponticums near water at Lower Bay and filled up with Dogwood.

Lower Drive. This road to Kingscote Station completed this year with hard flint and slag below, well rolled and shaped. During the past Winter and Autumn grubbed the whole of the ground on each side, so as to expose the new road through the wood fully to the sun. Planted a variety of Poplars and other forest trees on the higher and northern edge of the drive, and on both sides sowed Furze, Broom, and planted small plants of Sloe, Box, and

Mountain Pine, with a view to get a pretty foreground and yet not shade the road. This road has taken over two years to form. Fixed iron post marking date of making it. 1902

Steps and Paved Ascent to Bowling Green.

Formed, in the Spring of this year, the steps laid in earth with Rock-Flowers beneath them and the pathway, flagged with old London paving-stones, in a direction driven through the group of trees north of the House, so as to approach the bowling green from its east end, and also as leading easily to the walks above the House.

Pergola round West Garden.

Paved with old stones from Christ's Hospital, Newgate Street, E.C., dismantled this year. Finished August 7.

West Garden Pavement.

Finished paving the paths in the west garden with old London paving (York stone). Edged all the beds with stone at the same time, edging mostly with new York stone.

Dates to Woods.

Placed stout iron date pillars to woods planted or replanted by me. 'Replanted' means planting all the spaces hitherto occupied by underwood with stout saplings, mostly of native forest trees, tall enough to keep their heads above the underwood for a time, and eventually replace it. One more cutting of underwood—say at an interval of seven years, after the plantation of the saplings—should give the trees such an advantage that they should soon settle with the underwood.

Catch Pits.

Formed this year several small pools in the path of the streamlets feeding the lakes, to form drinking-places for cattle, and also help to keep the ponds clearer. If these catchpits are cleaned of sediment every year (in times of drought), the coarser sediment will be much lessened in the lakes.

Drinking-place, Sands Shaw.

Pushed back a good piece of the wood here to give ready access to drinking-place in streamlet, and with very good result, as it opened up the valley (seen from the lake) and also gave a good view of the House.

Fencing Drives.

Upper and lower drives fenced with Oak posts and rails in dangerous places. Done by own man (Brazier) from own Oak.

1902

Pergola from West Garden to Bowling Green. Formed this with 14-inch pillars of local brick and cross-beams of our own Oak, it being intended for Wistarias, and therefore strong, and made so that from it the climbers might rise to the trees around.

Summer-house on Bowling Green. Formed this with local brick, and roofed with old Horsham stone left from roofing the House. Finished in November. Berceau beside this formed the following Spring.

Ride through Warren's Wood. Formed, with much labour and care in moving earth, a ride 18 ft. wide through the lower side of Warren's Wood, the object being to get an airy ride from the House to the Home Farm through the wood, avoiding the deep gorges in the same, and connecting easily with the rides in the upper part of Warren's Wood.

Old Quarry Top of Warren's Wood. This rough corner grubbed, fenced, and wired, to fit it to plant with Silver Fir, Japanese Larch, and Cedar.

Mine Pit Wood and 'Shaws' near. Planted here (December) 1,000 young sapling Oaks, 8 to 10 feet high, with the aim of supplanting underwood by trees. The trees came from Oudenbosch, in Holland. Although these Oak trees were large for young trees, they were not staked as usual; but all the side branchlets were taken off, so that the wind might have no power on the trees until rooted, the tree itself, as it were, forming the stake—a plan which had answered well in previous years.

1903

White Spruce. Planted a number of these in Kingscote Wood, near railway, small plants that came from Denmark, rather late in April.

Standard Pear Trees. Planted many of these in new orchard in the hope that Pears may bear well in this form.

Green Pathway round Lake. Made a convenient 7-foot grass pathway all round the Lower Lake, making effects of water better, and of the plantings too.

Spanish Broom and Furze. Sowed much of these seeds on the railway banks going through the estate: also seed of the False Acacia on the same banks and in some of the young woods.

Cornflower. Sowed several pounds of this on the railway banks between West Hoathly and Kingscote, in the hope of naturalizing it on the stony and shaly banks.

Alpine Laburnum. Pricked in a lot of this tree on the railway between the two stations, Kingscote and West Hoathly, in the hope of their beauty being some day visible thereon.

Shaws replanted. This Autumn replanted two important shaws that I grubbed some years ago. I made a mistake in clearing the very rough ground, impossible almost to get into a level or good state—for pasture, proving that the old people who planted them or let them get covered with trees were right. In up-and-down country grubbing shaws and small woods in hollows is very doubtful work, as I find to my cost. Prepared for replanting shaw at bottom of Crabtree Field, west of Moat, and shaw at lower side of the Knolls Field. Finished in December. Planted 1,000 healthy sapling Oak, 8 to 10 feet high, with fine roots.

Bedford and Huntingdon Willows. Planted 500 of each of these tree Willows in wet spots by shaws and by pools and streams.

1904–5

Cedar Grove. Planted on top of High Field (north of House) about two acres of Atlantic Cedars, Western Hemlock, Carolina Hemlock, Rocky Mountain Spruce, and Numidian Spruce. The ground had been prepared for a kitchen garden, but being too far from the House, was planted with these beautiful Evergreen trees. Finished in January.

Alders Wood and Gill below it. The corners of the near fields running into this in an awkward way, I planted them up and banked and fenced good lines round each field, thus throwing some rough ground into the wood. Extended the ride from this through the gill, and in the same gill planted 100 English Yews out of reach of cattle, and likely to give good shelter for birds.

Stone Dam to Lower Lake. The wall here having several times leaked, put a wall of rough stone inside old wall, puddling between the two walls. It is extremely difficult to keep such dams sound, and this has been giving way for ages. Put a mesh of close wire about the waterline over the clay to prevent the work of rats, and planted tall Reeds outside the rough wall to break the wash of the water.

Bowling Green. Commenced putting a low stone wall round this, as the best means of enclosing it, having previously tried fencing with a not very good result. The stone dug in the place at the Beeches quarry. Finished in July.

Narcissus planting. Planted over 50,000 Narcissus of various sorts in the pasture woods and orchard during the season now ending—the last planted late in March.

Sheds for Garden use. Erected iron-roofed shed for work in bad weather, 45 ft. by 20, to have an upper story for boxes and hampers and other stores. This shed completes and closes in the little working yard—formerly on my coming a deep pit hole—one of the 'iron-digging' holes, I suppose. One motive for making this shed is to have a good fruit-room kept at an equable temperature, neither hot nor cold, and without windows, and cut off from the south as much as possible. Also a root-room, and for the keeping in winter of some kinds of vegetables. Road to these sheds remade, and bottomed well at same time.

1904 — *Sheds for Garden use.*

1905

Formed new vegetable garden east of House, trenched and made road, intending to leave the upper walled garden for fruit. To be called the New Orchard in distinction from the Old Orchard east of House. Added to this garden in December, and made terrace of earth to cut off orchard from slope below.

New Orchard.

Took up turf here and made a garden of Evergreen shrubs beneath the Apple trees, planting herbaceous plants between, taking advantage of the light and shade of the trees to heighten the effect of the flowers.

Old Orchard east of House.

FLOWER GARDEN IN PINK AND ROSE TIME

1906–7

Little Stockards Field. Planted this with many thousand native and other trees—Pitch Pines, and Riga Firs, and wired against rabbits. It was a poor sour field and no good for pasture. Also planted many thousand trees by the rides and in the woods.

Yellow Pine (Pinus ponderosa). Planted two thousand little plants of these on the rocks in Mill Place, fencing and wiring the ground with care, as it was very difficult to get the rabbits out of the rocks.*

Stone Piers and Wrought-Iron Gates. The way to the House not being clearly marked by the fence and gate which were formerly here, erected these piers and gates from designs by Mr. Ernest George, R.A., using our own stone; erected by Jenner and Grynyer of East Grinstead.

1907

Forecourt. Formed this and built a wall round it from a design by Mr. Ernest George. Before this the vegetation came down too close to the House. Having found the forecourt, such as it was, without any proper stone foundation, removed what was there, and put in three hundred loads of stone bottom. Finished in February.

Grass Terrace above South Lawn. The slope being too 'sharp' here formed this so as to be nearly level, and removed broad border, which was never satisfactory. Put low walls at each end of grass terrace.

Blacklands Farm and Woods. Bought this at sale of the Courtlands estate as it lay alongside, and makes a good addition to the estate, bringing it up to over a thousand acres in extent.

---

* NOTE, 1910. *The rabbits beat us in the end and devoured the whole lot.*

1908

Pergola for Flower Garden. Remade this using stout Oak pillars instead of Larch, with which I first made it and which decayed very soon. Charred the Oak posts, hoping to make them endure longer.

Mixed Border. Formed this in direct continuation from the pergola, as we had no good one before. Dug the shaly rock out and made a border nearly 3 ft. deep, intending it for the choicer perennials. Backed it with a post-and-rail fence and trellis for Climbing Roses and Clematis.

Down Fireplace for Wood Fires and Chimney. Rebuilt the hall chimney from the base to the top, as the space in it was not sufficient to allow of good down fires of wood. Made the chimney 14 inches square all the way up, with the result that we get perfect draught and good fires both in hall and book room.

Peach Walls. Finished walls projecting from the main wall of fruit garden and planted them mainly with Peaches. These walls were planned to cut off the east wind and give us a warm place in the parts between the walls. Gave a good coping to each, and in the early spring put up deep wooden coping. Hitherto we have failed with Peaches owing to the high ground and exposure, and hope this new plan will succeed.

Planting Maidenhair Trees. Cleared the old quarry at the end of the lower lake of Briers and Withies, which choked it up. Planted it with nice clean young trees of the Maidenhair Tree in the hope that the shelter and soil would encourage a good growth. About thirty mostly strong, well-grown trees were planted 9 to 12 ft. high. Also put many of the roots of the early Bayonne Narcissus in the same place.

Planting Stone and South Fields. Finished planting in March the two large fields near Kingscote Station, Stone and South Fields, with trees mostly known to be hardy, and those not usually planted in masses in

1908

Planting Stone and South Fields.

woodland in the district, evergreen and summer-leafing, with a grove of Evergreen Oaks between the two fields where the old dividing fence was. Most of the plants were very small, and when planted not more than an inch high. All the fields were carefully wired at first against rabbits. No preparation for the trees whatever was given to the field, which had been in arable cultivation for many years.

HEARTH IN DINING-ROOM

WATER AND WATER-SIDE PLANTING

The greatest gain of our own day is the coming of the beautiful Water-lilies, hardy enough to endure the waters of our country. There is, perhaps, no land so blest with many waters as our own. In estuaries, ponds, lakes, we are rich, but few take advantage of them for flower culture, though nothing gives back in pleasure more than thoughtful water and water-side planting.

Between the water plants and the many useful and beautiful water-side plants, and those that may be grown on the banks near, we have rich resources in things that do not need much attention as regards cultivation, but will look after themselves in the grass, and even the coarse vegetation that grows naturally by some water-sides. As I have had much experience with this phase of hardy plant life, I give a brief review of my own ponds as described in *Country Life* of March 6, 1909:—

'The ponds, when I saw mine first, had no vegetation round their margins, but a hard, yellowish line above the water, the soil being a stony, poor loam with a sandstone bottom. The first thing to do was to abolish the ugly bare line, which happily we had ample means of doing. Planting of the near margin of the water gave us many opportunities of using beautiful plants—Reeds, Loosestrife, Iris, and Bulrushes. These in our country give us as good effects as any, as one may see in natural lakes like the Broads and the backwaters of our rivers. But while we should make sure of getting all our own native water and water-side plants, we find good ones in the flora of other countries, and these we use. The Water Iris, the common native Reed, of which there is a larger variety, the Reed-mace

Water and Water-side Planting.

(*typha*, three or four), and the Water grasses. The very best of the hardy Bamboos associate well with water-side plants, and all the better if in bold groups; or if not exactly at the water-side, very near it, the best and most striking for Winter effects being the palmate Bamboo and the common Japanese Bamboo (*metake*), and two or three other hardy kinds. The dwarf kinds are of no use, not giving as good effects as some native Reeds.

Pampas grass, though it fares ill in cold districts, is a great plant grouped near, though not quite on, the water-side, the autumnal effects of it being fine, and all the year round it goes well with Reeds and other water-side plants. In Northern districts it might be too tender, and it is a great thing to deal in the neighbourhood of water only with hardy things. Among Spiraeas, the plant called Goat's-beard goes very well with the tall and vigorous Giant Spiraea and other herbs. Some of the taller Money-worts are pretty and effective, and they are rarely seen in gardens, especially the species *vulgaris* and *thyrsoides*. Aconites, handsome but dangerous plants in gardens, often do well by water, and the blue dandelion-like plants (*mulgedium*), which are troublesome in gardens, are also pretty in a similar situation. Sometimes I put the tall Moon-daisy among a delicately-coloured Aster (*puniceus*), and, both being tall plants, together they have a charming effect in the Autumn. Golden Rods put beside a pond do very well, and struggle with their heads well up among other growths, though they are not true water-side plants. I am very fond of the true Bulrush and the flowering Rush, which always group themselves in pretty ways. These are among the plants that help one to soften the margin of a lake. Our large native Water-dock is a stately plant, often growing naturally by rivers, but not always to be found in nurseries, and it too must be sought beside water. The Arrow-heads are very graceful in effect of foliage and also flower in a pretty way, the most attractive to me being the

LOWER LAKE (SUMMER)

Water and Water-side Planting.

single ones; and both single and double run about and increase rapidly, disappearing altogether in Winter. The Royal Fern is essential, though it does not thrive so well in loam as in peaty soils. Certain of the bolder herbaceous plants of Northern Asia, which are impossible in the garden owing to their vigour, come in here very well, where they may be allowed to take their own—often wild—way. Thus the Giant Knotworts (*Polygonum*), which soon become a nuisance in the garden, are often good by water, and effective all the Winter even from the colour of their stems. The same may be said of other vigorous plants, such as the Willow-herb, both white and red forms.

It is worth noting that we have four sources of supply for our water planting—trees, among which are some of the most graceful; shrubs like our native Viburnum; water-side plants, of which there are a number; and true water plants like the Water-lily. These of late years give the crowning charm to the water itself; their beauty of colour is extraordinary, and they are almost worth growing for the sake of the foliage alone. Their hardiness in our climate is remarkable. Some groups have been eighteen years in one place, and though never touched or attended to in any way, have improved every year. Where they get too thick the number and size of the flowers lessen a little and, therefore, if there be time, it does good to thin them. They thrive in deep water, and big groups are to be seen on the shallow margins. One sowed itself in 12 feet of water and has come up many years in the same position. No frost seems to touch them, even when the ponds were emptied for cleaning; the only one touched was the yellow *cromatella.* The great buds of that certainly were hurt, but it recovered itself in a year and flowered as well as ever. The water is fully exposed to the winds and waves, so that the plants get no such shelter as they do in small sheltered pools. But while they will grow anywhere, one can never enjoy their full beauty unless some enemies are kept down. Chief among them are

Water and Water-side Planting.

the water-vole and the water-hen, the former carrying away the flowers and eating them at leisure, the water-hen disfiguring them as they grow. It is easy to keep these enemies away in small ponds, which perhaps they will not always frequent, but in larger areas of water it needs perseverance, and we find if we cease trapping that the flowers disappear. The water-vole is not so difficult to exterminate as the brown rat, but the water-hens come in a continual stream from the woods and copses, no matter what we do. The presence in any numbers of the ordinary collection of water-fowl, which we see in so many places, would disfigure the Lilies greatly and make their culture not worth attempting. These Lilies have never received any ordinary gardening attention in the way of special soil, but were planted simply in the pond mud. When cleaning the ponds we leave a little islet or bank of this mud here and there, and on that the plants thrive to perfection. First the roots were thrown in tied to pieces of brick or waste iron. Some of the kinds increase naturally from seed to a degree that, if neglected, would spoil the general effect, so we have to cut the seedlings out, or the water would become covered. It is essential to have open pieces of water among them and not to overdo—a common fault. In the Spring the effect of Narcissus is a great help, and some beauty from that source may be looked for throughout the months of March and April, and even a good deal in May. Most large kinds, like Emperor, thrive near the water and give fine effect.

Among the shrubs a few of the wild Roses help, particularly the American Marsh Rose, which is pretty both in flower and fruit, and comes in bloom after our own wild Roses. As regards the tall herbaceous plants, we find it gives a better effect not to cut down the stems in Winter. The tall Siberian Knotworts and Meadow-sweets and Reeds are as interesting to a plant-lover in Winter as in Summer, the seed-vessels of the Reeds, &c., showing all through, and unless compelled by changes or other reasons, it is better to let them alone and cut down in April. It will be noted that all the plants used are such as take care of themselves among Grass and other plants, as any hoeing or cleansing in the ordinary sense would destroy their effect; and only plants are used which do not mind this.'

Water and Water-side Planting.

STONE BENCH IN PLAYGROUND

THE HEATH GARDEN

The landscapes of the British Islands and Western Europe owe much to the Heaths; but these have been slow in coming in any effective way into gardens. It is true that many close gardens are not the best suited for them, and that they are happier where there is room to treat them in bolder ways than is easy in the small places. I began rather slowly and doubtfully with them, charmed at first by the little Alpine Forest Heath, and even that is generally so ill-grown that we seldom see the fine effects it may give. The next that charmed me were some of the bright forms of the Grey Heather (*Erica cineria*); and so I went on until one cold winter's day, when searching in a nursery for wholly different plants, I saw a beautiful little bush of rosy-tinted flowers, and this was the Portuguese Heath (*E. lusitanica*). That settled me as a Heath-lover, and I have planted as much as I could get of it since. For many years it has been a pleasure, as it is good enough to flower all the Winter and far into the Spring. Just now (in March) this plant is in great beauty, and with it in bloom is the little Alpine Forest Heath and Hybrid Heath, all in masses. Usually this fine bloom occurs a month or two earlier in the South of England, so that one may say that you may enjoy the beauty of the Heaths all the year round, and the effect of the Heath stems is always good. But the best effects are not to be got without massing and grouping, and therefore it is in gardens where there are poor banky places in which 'nothing else will grow', or on some spread of rough land, that we get their best effects. Rich soil is never wanted for them. We mostly begin with the idea that they want peat, and certainly they do very well in peaty and sandy soils; but

THE HEATH GARDEN (SUMMER)

in many cases they grow in loamy soil without a particle of peat. Even the charming Connemara Heath is quite happy in such soil. As regards their endurance, some of the Heaths will grow in anything or nothing. Near me there is a railway cutting, very high and facing due south and without much soil, as most of it was removed. I suppose the seeds of the Heath floated in, or how should they come there and cover it so beautifully? I have watched them in hot years like 1893, and they flowered in that position when there was no rain for months, and never shrank nor withered. A Heath garden need not be rocky or in any way pretentious, but quite simple, for Heaths do as well on level ground as on the moorland rocks. Though they grow best, perhaps, in peaty wastes, it would be a mistake to suppose that only such soils can grow Heaths well. Some kinds even flower better on loam than on peat. If rocky banks or large rock gardens already exist, the dwarfer Heaths may often form their best adornment; but such are by no means necessary. Their fine effects are best known to those who see them on moors in broad masses. These can hardly ever be shown in small gardens, but why should they not be in large ones? It is by no means necessary to have a garden to cultivate Heaths in a bold and picturesque way, as almost any rough open ground will do, and some kinds will do among bushes and in woody places. The larger Heaths, where grown, should be massed in visible groups, and the dwarf ones seen in dwarf masses also, and not treated as mere specks on 'rockeries', or used as edging plants only.

The Heath Garden.

Associates of the Heaths

Native or European plants which frequent like places are best, such as the Brooms and Furzes, particularly the little Furze that flowers in Autumn, some of the Dogwoods also, such as the red form of Florida, and one Japanese Dogwood (*Kousa*), and some of the dwarf Barberries and Butcher's Broom,

The Heath Garden.

and the dwarf Rhododendrons of the mountains of Europe if the soil suits them. We allow the native Grasses and other plants to grow among the Heaths, not attempting to hoe or clean as in a garden, except when a certain rare kind is newly planted, when the ground is kept open until it is established. If one has mossy stones about on planting, it helps in effect, and also cultivation, to place a few among the Heaths, especially when planting new kinds.

Alpine Forest Heath (E. carnea).—The most precious little mountain shrub ever introduced. On many ranges of Central Europe at rest in the snow in Winter, in our mild Winters it flowers in January in the South, and in all districts is in bloom in the dawn of Spring—deep, rosy flowers carpeting the ground and the leaves all good in colour. There are one or two varieties, one white. This Heath is not averse from loamy soils, and often thrives on them as well as on peat soil.

Bell Heather (E. Tetralix).—This beautiful Heath is frequent throughout the Northern as well as Western regions of Great Britain. It is easily cultivated, and, being dwarfer than most others, requires rather more open situations; it also thrives best in boggy places. It has several varieties, differing in colour mainly. *E. mackaiana* (Mackay's Heath) is thought to be a variety of the Bell Heather. There is also a supposed hybrid between this and the Dorset Heath. I do not find this do so well in loam as in moist peat.

Broom Heath (E. scoparia).—A tall and wiry-looking Heath, reaching 8 ft. or more, flowering in Summer, not showy. I have seen this in cold parts of France (Sologne), as well as more abundantly in the west and south. It is hardier than most of the larger Heaths. It is often naked at the bottom and very bushy and close at the top, and though not quite a garden beauty as others are, it is worth a place in the Heath garden.

Connemara Heath (E. Dabeoci).—A beautiful dwarf bush, the name of which has been so often changed by botanists that it is difficult to find it by name in books, and I give it by the Linnean name here. It is a beautiful shrub, 15 in. to 30 in. high, bearing crimson-purple blooms in drooping racemes. There is a white variety almost more beautiful, and one with purple and white flowers. It often flowers with me four months throughout the Summer and early Autumn. The usually seen purplish form of this is dull in colour, but there is a very bright and finely-coloured red form which is to be found in some nurseries and should be sought for and increased, as it is far finer in effect than the ordinary kind. Old plants of this which have been several years in strong bloom will often be the better for being cut down in the Winter to one-half their height or more.

Cornish Heath (E. vagans) is a vigorous bush worth growing as a low covert plant, thriving in almost any soil, and growing quickly to a height of 3 ft. or

4 ft. A native of Southern Britain and Ireland; there are several varieties, but they do not differ much from the wild plant, and there is a cream-coloured form; the best in effect is the red form, a little higher in colour than the wild one. An excellent plant for bold groups and masses; it grows so close that the weeds have not a chance with it, although it may be well to cut out an odd brier or two in Summer. *The Heath Garden.*

Corsican Heath (E. stricta) is a compact bush, quite hardy in England and Wales. It is one of the close-growing kinds, and keeps itself clean and free from weeds. The dried flowers often keep on the bush a long time and give a curious effect. It flowers in Summer and Autumn, and is sometimes known in gardens as *E. ramulosa.*

Dorset Heath (E. ciliaris) is a lovely plant, as pretty as any Heath of Britain or of Europe. A native of Western France and Spain, it also comes into Southern England, and we find it hardy much further north than the district it inhabits naturally. The flowers are of a soft purple crimson, and fade away to a pretty brown. It is excellent in every way, thriving also in loamy as well as in peaty soils, and is very pretty in Autumn, often even into late Autumn.

E. multiflora.—This is a French plant and a native of Southern Europe generally. It is quite hardy with us and very free with its white flowers, but planted near our own Cornish Heath it does not seem much more than a dwarfish variety of that; but it is a close-growing good sort for the Heath garden.

Heather (E. vulgaris).—As precious as any is our common Heather and its many varieties, none of them prettier than the common form, but worth having, excluding only the very dwarf forms. The Heather is excellent for forming low cover, and of all the plants none so quickly clothes a bare slope of shaly or rubbishy soil, not taking any notice of the hottest Summer in such situations. Among the best varieties are *Alporti*, *coccinea*, and *Hammondi.* The plant and its varieties are well able to keep their 'heads up' among weeds and eventually to suppress them. Some of the bronzy forms are distinct and excellent in Winter effect, but of course this will not be felt unless they are in sufficient mass. One called *cuprea* is very effective.

Hybrid Heath (E. hybrida), a cross between *E. carnea* and *E. mediterranea*, is a most useful plant for those who desire good Winter or early Spring effects in the open air. The Hybrid Heath thrives in loamy soil almost as well as in peat, and is quite easily increased by division. This good little plant was found by the late James Smith, of Darley Dale Nurseries, Derbyshire, growing between plots of *E. carnea* and *E. mediterranea.* It was not in flower, but seeing its distinctive character he took it away and increased it, and a very precious plant it is.

Maw's Heath (E. maweana).—This is one of the handsomest of the hardy Heaths, and was discovered by Mr. G. Maw, in Portugal, in 1872. It may be described as a very vigorous-growing variety of *E. ciliaris*, which it closely resembles, but it is more robust; the flowers also, besides being larger than those of *E. ciliaris*, are darker red in colour. It flowers from July to December.

The Heath Garden.

Mediterranean Heath (E. mediterranea).—A graceful, bushy kind, growing 3 ft. to 5 ft. high, best in peat, and flowering often very prettily in the Spring. Although a native of Southern Europe, it comes into Ireland in the western parts, and is little more hardy in our country generally than the tree Heaths of Southern Europe—is, in fact, hardy in Northern England. Of this kind there are several varieties. It is very pretty in effect in Irish and Western Coast gardens. It lives in loamy soils, but has not quite its best effect in them. The late Mr. Boswell Syme, one of the best of British botanists, used to think the Irish form of this Heath a distinct plant, and it would be interesting to grow it and compare.

Portuguese Heath (E. lusitanica).—This is, for our islands, the most beautiful of the taller Heaths, growing 3 ft. to 5 ft. high, and, being hardier than the tree Heath, it may be grown over a larger area. Even in cool districts we have had it in a loamy soil, and almost every year it bears lovely wreaths of flowers in Winter and early Spring, white flowers with a little touch of pink, in fine long fox-brush-like shoots. Very rarely it is cut down by frost, but it springs up again from the roots. It seems quite happy in the country round London, and grows in brown soil as well as in peat.

Scotch Heath (E. cineria).—A bright and pretty Heath in many parts of Britain, particularly northwards, and breaking into several pretty varieties, such as *bicolor, coccinea*, *rosea*, and others. Flowering in Summer in ordinary soil. The brightest-coloured variety is well worth increasing. It is not so strong-growing as other kinds, and therefore should be grouped apart.

Southern Heath (E. australis).—A lovely bush Heath of the sandy hills and wastes of Spain and Portugal, 2 ft. to 3 ft. high, flowering in late Spring in Britain. The flowers are rosy purple and fragrant. It is quite hardy in the Home Counties, but I do not know how far North. It is scarce in nurseries as yet, not submitting to the usual ways of propagation; but it must be sought for, as no Heath garden is complete without it.

Tree Heath (E. arborea).—A graceful shrub of Southern Europe, North Africa, and Madeira, white-flowered and covering vast areas in the upland woods of Evergreen oaks, under them attaining a height of 12 ft. or more in North Africa, and in the Canaries becoming a tree. The Heath is tender in Britain generally, but may be grown in southern and warm districts, and on warm soil in sheltered valleys near the sea with its friendly warmth. The effect of the tall white 'meadows' of it in flower in the mountains of Algeria is beautiful. There is a variety (*Alpina*) likely to grow where the usual form may be tender.

STREAMLET WITH ARROWHEADS

SPRING FLOWERS IN TURF

The Wild Garden means the adornment of the ground away from the garden. Writers who approach this subject from the architectural side suppose that it means a new phase of gardening round the house, whereas it has nothing to do with that. It may be carried out in all sorts of positions, such as rough banks, old lawns, orchards, hollow ways and barren ground, or any suitable places in woods or along woodland rides, in any turfy or woodland situation outside the garden and away from it altogether. The beauty of the Wild Garden is that we get away from the need of the ceaseless care of ordinary gardening. If we put the things in likely soil and situations there is no more trouble in weeding, cleaning, and the usual work of the garden. Another point is that, if we make the best of it, we get in every way the charming effect of a Spring garden, more than it is possible to do in the garden proper. We get all the flowers of the Spring that our soil can grow, without interfering with the Flower Garden at all. A great advantage this, because it enables the gardener to deal with his Flower Garden in the best way. Where the Flower Garden was in the past, and still is, round Paris and many places, rooted up every year for the accommodation of Hyacinths and Wallflowers and plants of that kind, it is impossible to give any attention to the true Flower Garden, such as one ought to have near the House, that is, for the cultivation of the choicest Summer flowers — Roses and Carnations and favourite plants, which may want good cultivation. To make a true Flower Garden we ought to be free to plant all the Autumn and Winter as we wish, and if the ground is free for us there is nothing to stop this; and if we take advantage of

Spring Flowers in Turf.

the Wild Garden there is nothing to prevent us from doing as we please with the garden proper.

A great many plants are suitable for Wild Gardening, especially the stronger growers, and, in the Autumn, we have a number which are pretty by the water. The result is that we can enjoy some of the charms of the Wild Garden all the year round. But the greatest move of all has been in the direction of the flowers of Bulbous plants, which, in our country, so often do well in the grass. The most precious of all these is the Narcissus, for its hardiness, beauty, and vigour in our islands. A native of the great uplands of Europe, and very often flowering much later there than it does with us in Spring, it takes to our climate as no other plant does. Its effect on the landscape is excellent, and it does well for cutting for the house. Many beautiful kinds of Narcissus have been raised of recent years. Very often these are expensive, and will be so for some time; but all the good effects of which the family is capable can be realized in the Wild Garden with kinds which are by no means expensive. Moreover, we must mass and group them if we are to get any good effects. They help to give us a beautiful Spring garden, which in many cases we can enjoy from the last days in February until some time in early June. They have a vigour of constitution which enables them to grow in most unlikely places, and cold and wet soil is no disadvantage to them. Some kinds which seem to do poorly in gardens do well on clay banks and in other unpromising situations. I have planted many among Heaths, and they looked quite beautiful among the brown stems in Spring, as they do among the brown bents of the long grass at that season.

Overdoing.—One common error deserves to be guarded against, and that is dotting them over the ground regularly as tile-work. We see the results of this even in important public gardens—no relief, no verdure, no grace of arrangement. A hundred Daffodils on one side of a green patch may be more

HOUSE FROM SOUTH LAWN (SPRING)

Spring Flowers in Turf.

effective than all the turf dotted over with them. And wherever a group is placed it ought to have a natural and picturesque outline; it should not be evenly dotted. The roots should be set rather thicker towards the centre of a group or mass and scattered towards the margins.

Quite twenty years ago I got hold of a few sacks of roots and put them on a clayish slope planted with Cider Apples, when a friend said, 'You do not expect them to flower in that coarse grass?' and every year since they have flowered beautifully, with no trouble in planting beyond turning up a sod and putting a few Bulbs under it. We group them in natural masses along woodland walks where the soil is cool or heavy, and they do admirably. The places where they do not flower well are those with hot and dry soil. The best time for planting is the Autumn, and the earlier the better if one gets good stocks.

These thoughts mostly concern what is called ordinary soil—brown loam and clayish; but in places one may come upon deposits of a quite different soil which may favour a wholly different kind of vegetation. In such places one may indulge in things which would have very slight chances in ordinary soils. Hollows of peat and leaf-soil are excellent for this, and one of the most charming gardens was made by Sir Henry Yorke, who found a rich deposit of leaf and peaty soil in his wood. In such soils Lilies and other things, such as the American Lily and Bloodroot, that disappear in ordinary soil, may be grown with advantage. Such deposits vary infinitely, so one can hardly tell what could be done without knowing the soil. Where bog soil occurs naturally it also may be used effectively for certain plants that love such soils.

Mixtures.—Trade and growers now offer quantities for this purpose, but the most effective way is to get each kind separately, as then we can adapt them better to the situation. The early

Spring Flowers in Turf.

Bayonne Daffodil, for instance, which is a pretty thing indoors or out, is best on warm slopes; the great Irish wild Daffodil will grow anywhere in rank soil and grass. The early kinds, like the Tenby and Henry Irving, might grow on the sunny sides of banks, which would make them a little earlier. No part of the plan should be planting them in mown or lawn grass which has to be frequently cut. They look much better and do much better in grass which is only mown once in the Summer if mown at all. Manure or compost of any kind is not in the least necessary for them in cool soils.

After the Narcissi, which give such beautiful effects in all sorts of situations, the other plants become less important for effect; but they also have likings as regards soil. The Snowdrop, for instance, hardy and free as it is, and delightful in many situations, in some stiff soils often goes back, living, but not doing well. In many soils it is a joy to see it covering the ground. The Crocus, also, is a little slow in some soils, and seems to do better in those of a limy or chalky nature, as on Colonel Baskerville's land at Crowsley Park in Oxfordshire. The effect of the Crocus, where it does well, ought to be secured, and it often does very well in grass. The beautiful Neapolitan Crocus Imperati grows quite freely in grass, flowering very early, and no doubt many others would if fairly tried. There are a number of Crocuses, many of which have not been given a fair trial in this way in our country.

The Grape Hyacinths give one of the prettiest effects in grass, and seen near Daffodils often give lovely colour in April.

Perhaps of all the blue flowers for naturalizing in early Spring the most effective are the Anemones, particularly the azure blue one called *Robinsoni*, which grows freely in any kind of soil; also the Apennine Anemone, which is very free

Spring Flowers in Turf.

in limestone soils and runs about everywhere, as in Ireland, but does not last so long in flower. The most charming in that way is the Greek Anemone *blanda*, which is an early and patient bloomer. I have not been so lucky with this in strong grass, but in light poor banks it does well. What we want is things which can be naturalized easily in grass. In certain soils one might try some of the Crown Anemones, but in some soils they are not quite hardy, and we lose them now and then, whereas the scarlet Anemone is quite hardy and thrives in grass where it fails in garden soil after a year or so. A very pretty effect it gives in grass too. The blue *Hepatica* of the mountains of Europe is charming—best of all the wild kind and not any variety of it; but some of its beauty is lost if it is grown in coarse grass, being almost evergreen, and therefore best as a copse plant where the grass is poor and thin. The Pasque-flower is very charming in a wild state in Northern France and England, and is easily grown, and might be naturalized. As it is grown in great quantities by the Dutch there is no reason why bold experiments should not be made with it, especially in chalky places which it likes.

Among the most beautiful flowers of the world is the St. Bruno's Lily, so beautiful in Alpine pastures. Undoubtedly we may naturalize it with satisfaction if we have good plants and put them in places not mown early in the year. The Camassia, a North-West American plant, bears a pretty purplish flower, does very well in grass, and looks pretty there. In a garden it is rather too short-lived in flower. The Snow Glories, by their hardiness generally and freedom in multiplication, should be worth trying, and would suit some soils. The Meadow Saffron, often called Autumn Crocus, is very free in grass, and some of its varieties are very handsome, but they lack the charm of Spring. The Winter Aconite, of which there are now two kinds, is very free and welcome, but does best in

Spring Flowers in Turf.

limestone, dying out in cold soils. Dog's-tooth Violet, especially the European kind, is very free in light soil, and very pretty too, both in leaf and flower. The American kinds of this, which are so interesting and beautiful, are still too scarce to try much in this way, though in peaty or leafy soils they would do well.

Snake's-heads, the best of which is our beautiful native one, which has such nice varied colour, are excellent in alluvial fields. The wild Gladioli of Europe and Northern Africa are very pretty in grass, though in some soils they slowly die out. The Day Lilies are splendid, and though more perennials than bulbs, the effect they give in grass is quite charming. In gardens they are too short-lived in bloom. Three or four kinds of them I have grown, and they never failed to give a good effect, and very often they grow so freely in gardens that they might very well be turned out into rough, grassy places. Snowflakes (*Leucojum*) are excellent, the Summer one doing best in deep soil near water, but the Spring one is best and freest in peaty or free soil. Lilies are peculiar in one way, that one cannot be sure of growing more than two or three kinds in one soil; but where there are deposits of leafy soil, or bog, one can grow the American Swamp Lily, and also the Tiger Lily. In certain soils one might try the Turk's-cap and the White Lily. Among the newer bulbs the Montbretias are surprising, sometimes, for vigour even in poor soils. One is justified in taking up and dividing them when new, but I have had them among shrubs for several years, and they seem to give the same effect every year, so that I think the kinds that are free and plentiful might be worth trying. The beautiful Bloodroot of North America should grow freely enough, yet it fails in some soils, though any one with natural boggy soil should welcome it. Among the Scillas the Spanish kind is a very bold plant, and worth growing in free, warm soil, and is as handsome as our own Wood Hyacinth. There are various other bulbs which may have some beauty, but are not free or plentiful enough to try in the bold way that Wild Gardening demands. Among these are the beautiful Wood Lilies (*Trillium*), which in many Southern gardens disappear slowly, but in Western and cool places live, and are so beautiful.

Spring Flowers in Turf.

THE WOODLAND GARDEN

Where the woodland comes near the house, or within easy access to it, one of the most interesting things is to take advantage of the beauty of native flower or shrub and add to its attractions in simple ways. This can be easily done without interfering with the growth of the trees overhead. In the Woodland Garden the first and main thing to do is to bring into view the beautiful wild plants that grow in the place; for one reason (among others) that nothing we shall probably ever plant of exotic things will equal their beauty.

A natural amphitheatre of old woodland was for ages paved with Primroses, but, densely planted with underwood in the days when underwood was profitable, the Primroses were only visible every ten years or so as the underwood was cleared. Later the wood was planted with tall saplings of forest trees, and much was done to get rid of the underwood, so that every year the Primroses were clear in view, and for three months it was a more beautiful sight than any artificial planting would lead to. So it is in the time of Wood Hyacinths and any of our native things that are worth seeing. Slight differences of soil, elevation, rainfall, and other conditions give us different and often beautiful effects, and our first care should be to find out what thrives and enjoys the conditions of the place.

After our own native flowers certain others deserve a place in the Woodland Garden, paying heed always to the nature of the soil. All the now great family of the Narcissus—many as hardy as our own wild kind—are a great source of beauty in all moist and cool soils, and are, moreover, able to take care of themselves as well as native weeds. Twenty years ago, I put in some sackfuls of Narcissi in a stiff bottom full of Cocksfoot grass. Friends said they would never fight the long grass, but they have grown and have flowered well every year since —the flowers often over 20 in. high—and neither the Cocksfoot nor anything else has injured them. Also, at the same time, we planted in a bank of Heather some vigorous kinds, and they have never gone back and never failed to flower and show among the brown-hued Heath in Spring. In some light soils Narcissi die back and never flower, and it is useless to persevere with them; but the very soil which is so unkind to them may give us good effects of Crocus, Snowdrop, or Scilla. About the Snowdrop it is curious to note that in some soils it runs freely like Twitch and gives beautiful effects, while in others, no matter what is done for it, it gets weaker and often dies out. No knowledge of soils or long words about them will help us so much as experiment, and even native plants die off in certain soils.

The Woodland Garden.

We find that certain plants, too vigorous for the garden, are not so easy to get rid of, and these may now and then be well placed in the Woodland Garden. Among such plants are the greater Knotworts, the larger Starworts, Cow Parsnips, the Bee Balm of the Canadian woods, Goat's Beard, and some of the Spiraeas. All, however, depend for effect on soil and situation.

One of the most amusing things in the Woodland Garden is the establishment of native plants absent from the district. In boggy spots in woods the most stately of all plants is the Royal Fern, which has been exterminated in a large area of Southern England where it was once common. Solomon's Seal is not a common native plant, and it is graceful and well worth introducing, and so of the Lily of the Valley in sheets in some woods, and the Winter Aconite, in warm and chalky soils. We must not forget that even wild things cannot be well established if the soil is against them. The common water Forget-me-not, among the prettiest of Summer flowers, is very easily established where it is absent in streams or pools. Where in the North, or

The Woodland Garden.

in uplands, the woods have peaty or free open soils, we can establish other things of a usually different nature from what we get in heavy ground, and may even be happy enough to establish the Twinflower and the Winter Greens and Cyclamens.

Seed.—By the simple means of sowing seed something may be done, especially at the fringes of woods, or in old quarries, sharp slopes, or old fence banks. The three Furzes—the common, the tall Irish and the small Autumn-flowering—and also the common and the Spanish Brooms, are all easily raised from seed thrown out of hand, not covered or taken any further care of. The seed should be sown rather late in Spring or in early Summer—the best months are May and early June—and thrown broadcast, raising the hand high so as to spread the seed and leaving the rest to Nature. Where the Foxglove is absent it is easy to establish it by sowing the seed as soon as it can be gathered after ripening in the Autumn, or, failing good seed then, in the Spring, not too early, say April and May, taking small pinches of the seed and throwing it high so that the wind will help to spread it. I raised a handsome regiment of Foxglove along a ride in a Chestnut wood by sowing 1 lb. of the seed. This is only worth doing where the wild plant is absent. A sprinkling of the white and spotted kinds among the red is desirable. The seeds of all these things are saved in quantity by all the great seed-growers, and there should be no difficulty in obtaining good seed of any of them.

Evergreens.—After the joy of the Woodland flowers is past there remains to us the most enduring and, on the whole, precious charm in the presence of that growth of our native and other Evergreens which is denied to a vast area of country in Central Europe and North America. There is nothing to prevent our having the full enjoyment of Evergreens without detriment to the trees overhead. We may see a sheet of the tiresome Pontic Rhododendron as underwood, but we can do better by using our native Evergreens, of which the queen is the

The Woodland Garden.

Holly. If it comes naturally so much the better, but if not it is worth planting in the near home woodland; not everywhere, but in sufficient groups to be effective. Holly does differently in different soils, but it is worth having on all. It is often lost through the attacks of ground game, especially in severe Winters. Always use for the Woodland seedling plants of the wild sort not over 3 ft. to 4 ft. high. Any grafted or variegated or named kinds would be useless in the case, and not half as pretty in effect as the wild tree.

Next to Holly I should put the native Box, the beauty of which is obscured in gardens. Box is an Evergreen of supreme importance in Woodlands, in sandy, chalky, or bluffy places most of all. Its odour and colour are most welcome. In planting Box always use seedling plants, no matter how small, of the Tree Box. The garden forms are common in nurseries, but for the tree kind we shall have to go to a forest nursery. We call it the Tree Box to distinguish it from various dwarf and other forms used in gardens. The garden forms will grow in a stubby sort of way, but they will never have the plumy grace of the Tree Box grown from seed. It may be planted in rather dense Woodland without fear of loss, the tiniest plants being respected by the rabbits.

Ivy, which some are in a hurry to cut off trees, should not be destroyed. It is the finest Evergreen climber of the North. Instead of destroying it we should put here and there at the base of trees little shoots of the more distinct kinds of Ivy, of which there are many—such as the Amoor, the bronzy kinds, and small-leaved Ivies. Do not trouble to plant, but simply lay the shoot with the few fibres we shall generally find on it at the base of the tree and put a stone or a clod of earth or piece of turf on it, leaving the point exposed, and it will soon make its way up the tree. Another good native Evergreen is Juniper, especially for sandy and calcareous soils. Among dwarf Evergreens, our native Daphne (*Pontica*) is worth having,

The Woodland Garden.

not only for its constant verdure, but also for its fragrance in Spring. Having done justice first of all to our Native Evergreens, the next thing is to consider any exotic ones that will equal them in hardiness, beauty, and effect. We cannot exclude the Rhododendrons, but we can avoid those that are grafted on the Pontic kind, as it will kill the graft in time and then spread itself everywhere. Kinds on their own roots, not budded nor grafted, should always be preferred. In woods in which the soil is of a peaty, leafy, or open nature this class of plant is all the more free, and in such soils we may enjoy that most beautiful of all flowering Evergreens, the Mountain Laurel or Kalmia. The great Partridge Berry of North-West America is a fine Evergreen as underwood, though not often found in nurseries, and the Evergreen Barberry (*Aquifolium*) from the same region is also excellent. In the Woodland Garden nothing exotic should be used that is not, like this Barberry, as hardy as our Native Evergreens, or which want any care or cultivation. Many people think, naturally enough, that it is not easy to get Rhododendrons on their own roots. This is so in many nurseries, but there are in at least several nurseries in Surrey acres of plants grown from layers, though they are more expensive, no doubt. In the same nurseries and in many others seedling Rhododendrons are to be had in the attempt to raise new varieties, and these, of which nice bushes may be reasonably bought, may safely be planted in Woodland. In the Derbyshire nurseries certain kinds are grown freely from layers, like Cunningham's White, and are excellent for coverts. Lastly, we may layer for ourselves favourite kinds. Apart from their beauty, a minor reason for putting Rhododendrons in Woodland near the house is that in some heavy wealden lands the kinds that are so poor in growth in open ground will grow very well indeed in the leafy soil and slight shade of the wood in the same district in which they fail in the open. It would be interesting to grow the wild forms of the North American

Rhododendron, and Azaleas, too, in case we have room for such refinement in planting. Even the Pontic kind is worth a place provided it is not everywhere. The Cherry Laurel, which is hacked into so many ugly and impossible shapes in gardens and sometimes by the acre cut down to a flat surface, is really a noble tree if allowed to take its natural shape in the wood, though never in great quantity. In planting it in cold and inland districts we should take the hardiest, i.e. the Caucasian and the Shipka forms.

The Woodland Garden.

WATER-LILY 'COLOSSEA': LIFE-SIZE IN OPEN WATER
(ENGRAVED FROM A PHOTOGRAPH BY G. A. CHAMPION)

GRAVETYE MANOR

The following account of Gravetye is taken from the *Sussex Archaeological Collections*, 1858, Vol. X, pp. 166, 167.

The old Elizabethan stone mansion of Gravetye, in the parish of West Hoathly, retains much of its ancient appearance, as when built at the close of the 16th century, with its broad, mullioned windows and terraced garden. Its series of tall dormer windows grotesquely breaking up into fragments, its ponderous cornice, and the Doric triglyphs supported on corbels, and upholding pinnacles at the angles, give the architecture a very peculiar character. The stone porch has the initials H. F. on it; and in the spandrils of the west door are the initials R. I. and K. I. all relating to the family of Infield, who for a brief period possessed the property. The ceiling of the hall is ornamented with a number of devices in stucco:—a ram's head erased; an oak leaf, with acorn, as a crest; a bunch of grapes with vine-leaves; a rose with leafy branch; a pelican vulning itself. An iron chimney-back, in the hall, exhibits again the same initials, '*R. I., an° Do.* 1598, *K. I.*'

The family name seems to have been gradually changed from Hanningfield to Infield. According to the information of W. Smith Ellis, Esq.—who has kindly supplied other notices—in the Subsidy Roll (printed in our Vol. IX, p. 87) the family of Infield was descended from the Haningfields, of Haningfield, Essex, one of whom bare for his arms, 'or, a chevron sable,' and was at the battle of Boroughbridge, in the time of Edward II. A knight's fee was held in 1350, in the manor of Sheffield, in Lingfield, Surrey, by William de Inningfield and Nicholas de Malmeyns; and in 1531, William de Inningfield died seized of a property there, leaving Adam his son and heir, aged twenty-four. We find the name again in 1437, when Elizabeth Inningfield released all her rights in Lingfield to William; and, in 1483, William de Inningfield held a tenement in Lingfield, called 'Le Lynde Place'.

At the death of James Infield, in 1596, it was proved that he died holding certain messuages at Lingfield. He appears to have been succeeded by Richard Infield, who built the house (Manning & Bray's *Surrey*, ii, pp. 282, 284, 351). It appears from the tombstones in West Hoathly Church, that Richard Infield married 'Catharine, daughter of the right worshipful Edward Culpeper, of Wakehurst, Kt.' She died Oct. 16, 1623, and he on March 11, 1624. 'Richard Infield, of Gravetye, gent.,' probably their son, appears to have married Catherine Compton, and their eldest daughter, Agnes, who died September, 1685, aged thirty-nine, seems to have carried the property to her husband Henry Faulconer, son of Infield Faulconer, Esq.; and this explains the initials of H. F. on the porch.

SOUTH PORCH OF HOUSE

Afterword

PETER Herbert wrote in his afterword for the 1984 edition of *Twenty Years' Work round an Old Manor House* that so much has been done to keep William Robinson's work and memory alive to delight and thrill the generations to come. Peter's restoration of the garden was a huge project, a passion project, as it has also been for us. Gravetye is a special place to us personally, so to have kept it at the very highest standard has been a privilege and we have been delighted to have been part of its history.

Since acquiring it in 2010, we have undertaken a number of restoration projects in the hotel itself and also in the garden. We wanted to ensure that the philosophy of great country house hospitality continued, and to maintain and nurture William Robinson's garden and legacy. We were very keen to keep Gravetye's heritage and not erase it. There is something magical about the manor and its gardens that comes from more than bricks and mortar. It is an intangible quality that arises, we believe, from all that has taken place here over the past 425 years.

The appointment of our head gardener, Tom Coward, in 2010, proved critical. Returning a garden to its full potential was a daunting prospect, but Tom proved himself equal to the task. Peter Herbert rescued the garden in 1958 when he created a hotel from the beautiful Elizabethan manor that William Robinson had owned and made into this unique place. After Peter retired in 2004, the garden suffered from lack of funds, so that is when we stepped in. The garden today is a testament to Tom's vision, embodying all that William Robinson represents with purpose and creativity in a modern context, and thereby allowing the garden to prosper in a way that we hope Robinson would have followed himself had he still been here. Hopefully he is looking down with approval at the wonderful spectacle that is Gravetye Manor Garden.

Every part of this painstaking restoration project has been carried out with a sensitivity that respects the past but looks to the future. In 2018 we added a glass-fronted dining room, enabling guests to really interact with the garden. Fortunately the planners agreed that the modern addition of a restaurant was the right way to put Gravetye into the twenty-first century in order for it to survive. We now enjoy unparalleled views of the garden whilst eating the produce from the amazing walled kitchen garden. We feel it truly captures the essence of the latest chapter in Gravetye's ever-evolving story.

—Elizabeth and Jeremy Hosking

William Robinson, 1838–1935

INDEX

Illustrations in Italics

Gravetye
WR

PHOTOGRAPHY CREDITS

Mark Bolton: 12 bottom, 30 bottom, 38, 40–41, 42, 43, 44–45, 46

Clive Nichols: 21 bottom, 22–23, 27 top, 27 bottom, 32–33, 34 top, 34 bottom

Jeremy Pelzer: 20 top

Claire Takacs: Cover image, 2–3, 4–5, 6, 8, 10–11, 12 top, 14–15, 16, 18–19, 24, 28–29, 30 top, 36–37

Pages 2–3 and 4–5: William Robinson's love for this gentle, rolling, wooded landscape is expressed in part by the text of this book, which details his thoughtful attention to its care and development. Throughout his life he bought adjacent properties, culminating in an estate of over 1000 acres with the manor and its gardens at its heart.

Frontispiece: *Kniphofia rooperi*, cannas, and *Salvia* 'Indigo Spires' continue the colour display in the flower garden deep into autumn.

Published in the United States
of America in 2024 by
Rizzoli International Publications, Inc.
300 Park Avenue South
New York, NY 10010
www.rizzoliusa.com

For Rizzoli International Publications, Inc.

PUBLISHER: Charles Miers
EDITOR: Stacee Gravelle Lawrence
PRODUCTION MANAGER: Maria Pia Gramaglia
MANAGING EDITOR: Lynn Scrabis

VISIT US ONLINE
Facebook.com/RizzoliNewYork
Twitter: @Rizzoli_Books
Instagram.com/RizzoliBooks
Youtube.com/user/RizzoliNY

2024 2025 2026 / 10 9 8 7 6 5 4 3 2 1
ISBN 978-0-8478-3767-0
Library of Congress Control Number: 2024935856

DESIGN: Susan Evans, Design per se

Printed in Italy